2023 Banff Pork Seminar

Advances in Pork Production Volume 34

Banff, Canada
10 – 12 January 2023

Editors:

Ashley Steeple
Ben Willing

ISBN: 978-1-7138-8885-7

Printed from e-media with permission by:

Curran Associates, Inc.
57 Morehouse Lane
Red Hook, NY 12571

Some format issues inherent in the e-media version may also appear in this print version.

Printed with permission by Curran Associates, Inc. (2024)

For permission requests, please contact University of Alberta
at the address below.

University of Alberta
Department of Agricultural, Food and Nutritional Science
4-10 Agriculture/Forestry Centre, University of Alberta
Edmonton, AB, Canada T6G 2P5

Phone: (780) 492-3651
Email: pork@ualberta.ca

www.banffpork.ca

Additional copies of this publication are available from:

Curran Associates, Inc.
57 Morehouse Lane
Red Hook, NY 12571 USA
Phone: 845-758-0400
Fax: 845-758-2633
Email: curran@proceedings.com
Web: www.proceedings.com

Advances in Pork Production

Volume 34, 2023

banff pork seminar

Advances in Pork Production

Volume 34

Proceedings of the 2023 Banff Pork Seminar

Co-operating Agencies

Alberta Pork

Alberta Agriculture and Forestry

University of Alberta

Published by:
University of Alberta
Department of Agricultural, Food and Nutritional Science

ISSN: 1489-1395

For information please contact our office at:

Telephone: (780) 492-3651
Email: pork@ualberta.ca
Website: www.banffpork.ca

Mailing Address:
Banff Pork Seminar
c/o Department of Agricultural, Food and Nutritional Science
4-10 Agriculture/Forestry Centre, University of Alberta
Edmonton, AB, Canada T6G 2P5

Preface

A Sustainable Pork Journey

First and foremost, a warm welcome to all participants this year, in full presence for Banff Pork Seminar 2023. Another year has gone by since some of us met in a smaller format for the Banff Seminar in January 2022. Although we were a small group here in Banff, your support has allowed the Banff Pork Seminar to stay alive and be back this year for an event as we use to know it. Sustainability is a pretty common theme in a lot of areas in our society and our industry is for sure well aware of what it means. The pork sector, like other sectors of agriculture, has evolved tremendously over the years. There has been a diversification of different ways to raise and produce pork and as an industry we have to embrace and support all of these. Being sustainable does for sure have a strong meaning toward the environmental aspect of our production. Although agriculture is still often blamed, we have to be proud of the progress we have made in terms of being much more efficient in producing food for the world. The input needed to produce 1 kg of pork today is not nearly comparable to what it was years ago. This is something we need to talk about and let people know. Although, there is still progress to be made, we are for sure committed to always do better. Sustainability also mean being economically viable and this applies to all the sectors in our industry. Since spring 2020, there has been a lot of challenges. 2022 has been a year where feed costs have remained high, while prices have been good for the most part of the year. The packing industry has still seen a lot of challenges, which lead to more pressure on the entire pork sector. We need to have all factors of our industry do well if we want to remain strong and sustainable. I know that it is impossible to have a perfect economics alignment of all sectors, but by working more closely together, we should be able to do better. As always, this year's seminar will cover many areas of our industry. The economic picture, the environmental aspect, new technologies and others. Mostly, we wanted to be closer to the producer and farm by presenting subjects that would matter to the day-to-day work in the farm. The committee worked hard to find quality speakers from around the globe which can bring value to our participants. As always, we are thankful for the continuous support of our sponsors which year after year are here to help us to provide a quality program and event. Again, a big thanks to all the advisory committee members, to Ashley Steeple and her team and our two long time program directors, Ruurd Zijlstra and Michael Dyck. On my end this is my last year on the committee and as the advisory committee chair. I really enjoyed my time and the collaboration with the group. I'm sure the group will remain in good hands for the future.

Thanks again all for being here and enjoy your time at the seminar and in beautiful Banff.

Sincerely
Dan Bussieres
Chair, 2023 Banff Pork Seminar Advisory Committee.

2023 BPS Advisory Committee

Chair

Daniel Bussières, Groupe Cérès., Levis, QC

Program Directors

Ruurd Zijlstra, Professor, Swine and Carbohydrate Nutrition, Dept. of Agricultural, Food and Nutritional Science, University of Alberta, Edmonton, AB

Michael Dyck, Associate Professor, Reproductive Physiology/Biotechnology, Dept. of Agricultural, Food and Nutritional Science, University of Alberta, Edmonton, AB

University of Alberta

Ben Willing, Assistant Professor and Canada Research Chair in Microbiology of Nutrigenomics, Dept. of Agricultural, Food and Nutritional Science, University of Alberta, Edmonton, AB

Alberta Agriculture and Forestry

Jennifer Hannesson, Livestock Traceability Section, Edmonton, AB

Alberta Pork

Javier Bahamon, Quality Assurance Coordinator, Edmonton, AB

Industry Representatives

Eric Aubin, DNA Genetics, St-Lazare, Que.
Javier Bahamon, Alberta Pork, Edmonton, Alta.
Jason Brandt, Steve's Livestock Transport, Red Deer, Alta.
Maaike Campbell, Birnam Pork, Warwick Township, Ont.
Melissa Da Costa, Baseline Pork, Harley, Ont.
Steve Davies, Maple Leaf Foods, Carman, Man.
Jen Demare, Swine Heath Professionals, Steinbach, Man.
Heather Ferguson, Cargill Animal Nutrition, Puslinch, Ont.
Michel Larivière, PIC, St. Malo, Man.
Rick Prejet, Porcherie Lac du Onze, Notre Dame de Lourdes, Man.
Cordell Young, Precision Veterinary Services, Lethbridge, Alta.

Conference Coordinator

Ashley Steeple, University of Alberta, Edmonton, AB

We extend sincere thanks to the following companies and groups who have provided financial assistance to the 2023 Banff Pork Seminar and to the publication of these proceedings.

Sustaining Sponsors

Alberta Pork
Fast Genetics
PIC
Topigs Norsvin Canada

Premier Plus Sponsors

Alltech
Dept. of Agricultural, Food and Nutritional Science, University of Alberta
Hypor
Maple Leaf Foods
New Standard West Equipment
Olymel
Penner Farm Services & Western Ag Systems
Steve's Livestock Transport

Premier Sponsors

Alberta Animal Nutrition Lectureship
Canadian Pork Council
Cargill Animal Nutrition
CHR Hansen & Robinson Bioproducts Inc.
DSM Nutritional Products
HyLife
Jefo Nutrition

... *Sponsors of the 2023 Banff Pork Seminar* *(continued)*

Select Sponsors

Alliance Genetics Canada / DanBred Ltd.
Agribution Canada Ltd.
APC
Automated Production
Barkman Concrete
Boehringer Ingelheim Animal Health Canada
Dechra Veterinary Products
DNA Genetics
EastGen
Elanco Canada Ltd.
Farm Credit Canada
FeedFlo
Grand Manufacturing Co. Ltd.
h@ms Marketing
Hipra Animal Health
Innovative Heating Technologies
Jyga Technologies - Gestal
Kemin
Ketchum Manufacturing Inc.
Maximus
Merck Animal Health
Minitube Canada
Modern Veterinary Therapeutics
Neogen
Novus International, Inc.
Polar Pork Farms
South West Vets
Vereijken
VIDO
Virox Technologies Inc.
Zoetis

... Sponsors of the 2023 Banff Pork Seminar (continued)

Supporting Sponsors

Carlo Genetics
Le Porc Show
Nutrition Partners Inc. & Farmhouse Consultants
NutriQuest
OPTIfarm Ltd.
Purdue University

Index of Advertisers

... *Index of Advertisers* *(continued)*

Program of the 2023 Banff Pork Seminar
January 10-12, 2023

Plenary Sessions

Wednesday

- **Global Pork Markets: Update& Outlook**

  Brett Stuart, *Global AgriTrends*

- **Pork Production Trends & Outlook: Opportunities**

  Steve Weiss, *NutriQuest*

Thursday

- **Sustainable Food Production**
  Banks Baker*, PIC*

  Maple Leaf Foods: Our Sustainability Journey & Approach
  Mauricio Alanis*, Maple Leaf Foods*

- **Your Story Matters**
  Vincent ter Beek*, Fresh Air Farmer*

Morning Breakout Sessions

1. **Swine Health**

   - **Circovirus - A Practitioner's Perspective of Present & Future**

     Ryan Tenbergen, *Demeter Veterinary Services*

   - **Strep. Suis & E. Coli in the Nursery - A Health by Nutrition Approach**

     Brooke Smith, *Cargill Animal Health & Nutrition*

2. **Sow Management & Reproduction**

   - **Feeding the Sow & Managing the Farrowing Environment to Improve Nursery Pig Performance**

     Casey Neill, *Carthage Veterinary Services*

   - **What to Expect From an AI Dose & Best Practices On-Farm**

     Kyla Ripley, *Carlo Genetics*

3. **Technology**

   - **Digital Farm & Technology Insight**

     David Speller, *OPTIfarm Ltd.*

   - **Evaluation & Application of Technology in Swine Production Systems**

     David Rosero, *The Hanor Company*

4. **Innovators**

   - **Examining the impact of dietary nitrogen content and source on growth performance in growing pigs**

     Miranda Buchinski, *University of Saskatchewan*

   - **Kazachstania slooffiae alters immune system function and promotes an anerobic environment**

     Tausha Prisnee, *University of Alberta*

   - **Effects of feeding barley-based diets differing in fermentable starch & fibre and energy value on growth performance & organ weight in weaned pigs**

     Joaquin Sanchez-Zannatta, *University of Alberta*

   - **Effects of sow grouping practices on mixing aggression and production**

     Jessica Vehof, *University of Saskatchewan*

- **FX Aherne Award for Innovative Pork Production**

  Weaning Ramp by *Barak Doell & Robert Lafreniere*, *HyLife*

Afternoon Breakout Sessions

5. **Piglet Management**

   - **Colostrum Management & Piglet Survival**

     Kara Stewart, *Purdue University*

   - **Economics of Mortality - With a Look at Decision Tools to Help Improve it**

     Lee Schulz, *Iowa State University*

   - **Effect of Fostering Strategies on Piglet Performance**

     John Sonderman, *DNA Genetics*

6. **Grow-Finish Management**

   - **Late Finishers**

     Matt Ritter, *Provimi North America*

   - **Wean-Finish Efficiencies - The Often Overlooked Areas**

     Nat Stas, *PIC*

   - **Optimizing Finishing Feed Costs**

     Kurt Stoess, *HyLife*

7. **Feed & Nutrition**

   - **Reducing Feed Costs in Grow-Finish Pigs**

     Francesc Molist, *Schothorst Feed Research*

   - **Opportunities for Improving Feed Quality & Consistency while Reducing Feed Costs for On-Farm Feed Mills**

     Malachy Young, *Gowans Feed Consulting*

- **Feeding to Improve Sow Livability**
  Laura Greiner, *Iowa State University*

8. **Building Engineering**

- **Farm Building Energy Savings**
  Stephen Clarke

- **Swine Heaven in 2050: Building the Future of Pig Housing**
  Sebastien Fournel, *Laval University*

- **Manure Management & Treatment: State of Knowledge**
  Stephane Godbout, *IRDA*

Table of Contents

Plenary Sessions

Breakout Sessions

1. Swine Health

2. Sow Management & Reproduction

3. Technology

4. Innovators

5. Piglet Management

6. Grow-Finish Management

7. Feed & Nutrition

8. Building Engineering

TN70 MAKE YOUR BUSINESS A #1 HIT
THANK YOU TN70 FANS!
No extra labor needed, The TN70 nurses her own piglets
No 1. sow in the world
Brazil
Netherlands
Germany
Belgium
UK
Ireland
France
Russia
Hungary
Romania
Serbia
Czech Repu-blic
Croatia
Austria
Slovakia
Poland
Lithuania
Greece
Spain
Argentina
Uruguay
Costa Rica
Colombia
Italy
Canada
USA
Mexico
China
Japan
Philippines
Vietnam
Norway
Sweden
Finland
Denmark
Bolivia
Ecuador
Peru
Paraguay
South Africa
Ukraine
Portugal
tn70.topigsnorsvin.com
Topigs Norsvin

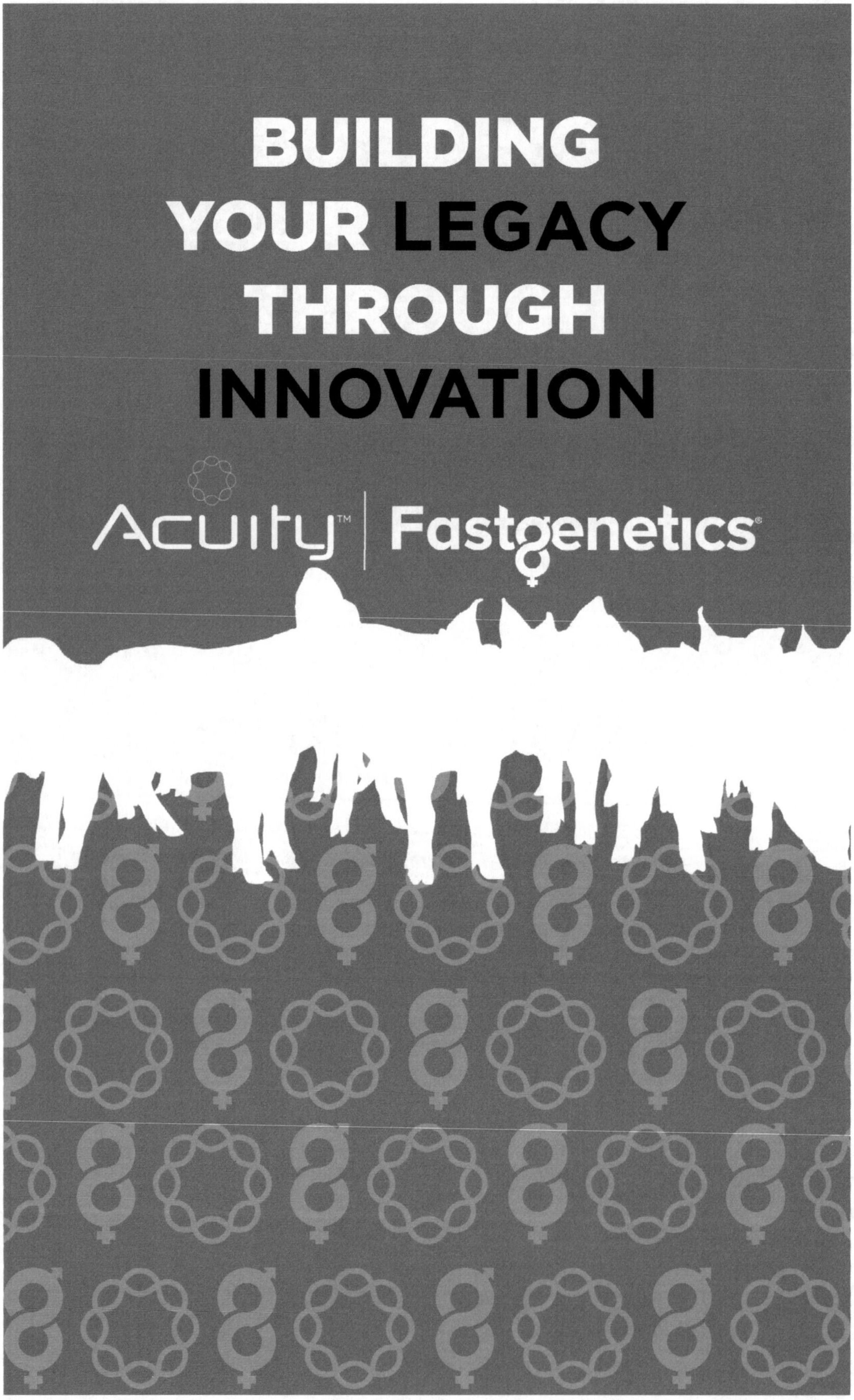
BUILDING
YOUR LEGACY
THROUGH
INNOVATION
Acuity™ | Fastgenetics®

PIC®
CAMBOROUGH®
YEAR OVER YEAR
SHE'S BETTER THAN THE
BEST
Never
Stop
Improving
Performance.
EFFICIENT | ROBUST | PROVEN
PIC has been improving the Camborough® for 60 years. That commitment has made her the world's highest-performing female in varying production environments across the globe. The Camborough® is easy to manage with fewer non-productive days and requires less feed in her lifetime than competitor sows. She continues to deliver heavier pigs with greater pre-wean survivability. What could be better? The PIC Camborough® next year. And the year after that.
Contact your PIC representative today.
PIC.com | 800-661-1543 | PIC.ca.info@genusplc.com
©PIC 2022.

Global Pork Markets: An Update and Outlook

Brett Stuart, MBA

President, Global AgriTrends

Global pork markets remain heavily affected by unprecedented financial stimulus, animal disease issues, and Russia's invasion of the Ukraine. 2023 takes us into uncharted water. Swine herds shrunk during 2022 in China, Europe, and North America. Those trends largely continue into 2023.

1. The Largest Macroeconomic Experiment in World History

As the COVID-19 virus began spreading from China globally in 2020, uncertainty and volatility spread. Lockdowns began all over the world seeking to prevent further spread from a largely unknown and lethal virus. Governments imposing lockdowns faced various degrees of pushback. Their response was the typical government response: more money. Political capital was maintained through massive and unprecedented stimulus payouts. And stimulus was needed to prevent serious economic damage. But the level of stimulus surged higher and higher. Within months of the pandemic, the G-20 nations had pledged US$5.1 trillion, but by May 2020 the total rose to US$9 trillion. By September 2020, the level of government financial support topped US$17 trillion[1]. These numbers continue to grow. In the U.S. the recently passed "Inflation Reduction Act" pledges spending and credits expected to exceed US$740 billion.

On a broad basis, assets are a function of money supply divided by the assets themselves. Massive, unprecedented spending led to money creation, spurring heavy inflation globally. The U.S. increased their money supply by +40% in 2 years. Raising that numerator in the equation led to surging broad inflation that remains today. While few voters will ever complain about "free" money, none can avoid the heavy impacts of inflation. 2023 begins with the economic hangover from 3 years of unprecedented government spending.

The inflationary impact from massive COVID stimulus drove commodities into a super-cycle. Drought in North America, and the Russian invasion of the Ukraine in February 2022 combined for a commodity surge trifecta. Feed grain prices soared higher and remain at historically high prices today.

This chart shows the average global prices of key food commodities. The recent commodity rally was extreme and has raised the global price of food.

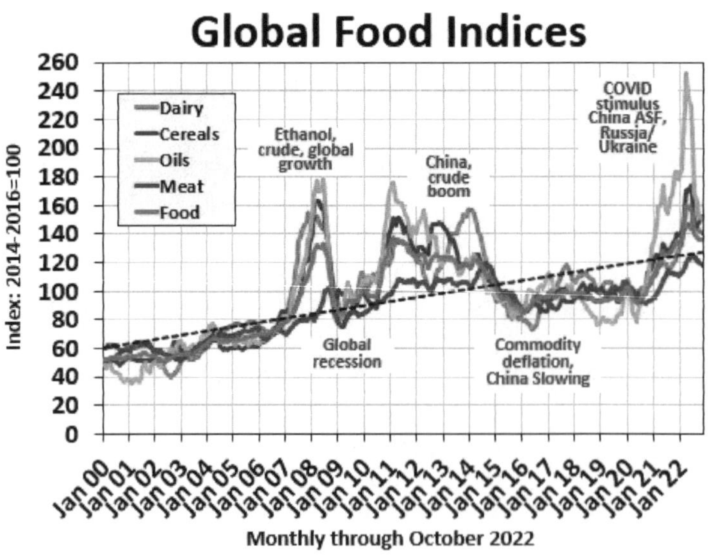

Figure 1. Global Food Price Indices (Source: UN/FAO)

The heavy stimulus also spurred consumer spending leading to demand surges in recent years.

Now central bankers around the world, led by the U.S. Federal Reserve, are scrambling to fight inflation. Interest rate hikes are most effective at slowing the velocity of money as a means to reign in inflation and have been widely used globally. However, weaker economies like Japan have not been able to match U.S. rate hikes, leaving their currency vulnerable to devaluation. The sharp yen declines against the U.S. dollar are shifting market shares toward EU pork at the expense of the U.S. exporters.

Rate hikes also portend recession. And recessions are beginning globally. However, in the U.S. pork at the retail price level has shown strong resiliency to recession. The depth and extent of recession in the U.S. and elsewhere lies in the hands of central banks.

2. Animal Diseases Spreading

Swine disease continue to roil global pork markets. In the U.S. and Canada, persistent battles with porcine respiratory and reproductive syndrome (PRRS) led to tighter hog supplies in 2022. USDA now estimates U.S. pork production down -2% in 2022, an unusual decline. This has been supportive to prices with the pork cutout maintaining high enough levels to hold margins above breakeven until last November. The reduced U.S. hog supply was able to buoy markets to profitable levels despite heavy feed cost inflation through much of 2022.

In Europe, continued ASF findings raised concerns across Germany with cases found near the French and Netherlands borders. Cases also spread in Italy, Poland, Hungary, Slovakia, Romania, Servia, Bulgaria, the Ukraine, and Latvia last year. The impact of disease restrictions has pushed Spain to the top of the export charts with those ASF afflicted nations relegated to exporting only within the EU. European producers have also reported challenges with the PRRS virus over the past year. Despite record high hog prices, European farmers report losses due to the sharp price hikes for feed, energy, and labor.

China's ongoing battle with ASF continues and will likely become a permanent fixture of the Chinese hog sector. With no viable vaccines and a massive swine sector involving tens of millions of small farms, the virus will likely remain for years. ASF remains in the Caribbean basin as well. U.S. Department of Agriculture (USDA) officials report efforts to surveil the afflicted herds in Haiti and the Dominican Republic, but also report that local officials report lack of funding and focus to work on eradication.

Fifty-nine countries have had cases of ASF in recent years. There is no commercially approved vaccine at this time.

3. China

China's hog and pork markets remain volatile in the aftermath of their ASF liquidation that began in 2019. The liquidation likely exceeded half of their total swine and led to shortages and hyper-inflation. Those high prices surged profits above US$300/head for 10 months in 2020 leading to widespread liquidation. Many farmers opted to breed market gilts in the race for profits. Those soaring profits led to widespread overproduction which caused price declines beginning in January 2021.

From January 2021 to June 2021, China's live hog prices fell by -66%. Margins turned negative in June 2021 and stayed there for most of the next 12 months. This led to liquidation and serious financial pressure on Chinese hog farmers. By June 2022, it was estimated that 20% of the herd had been liquidated and China's swine price "roller coaster" began again. Prices surged 108% by October 2022 and profits shifted higher. Rapid expansion, fueled by higher profits, remains underway in China today.

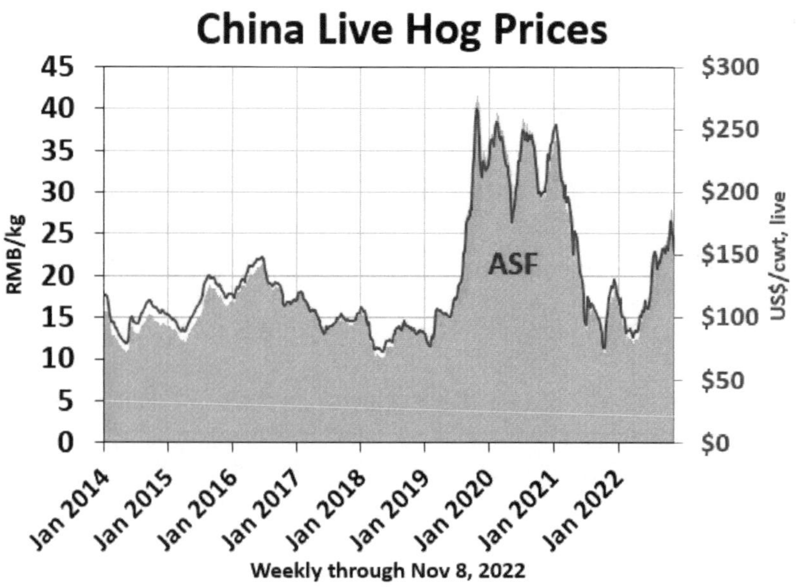

Figure 2. China live hog prices (Source, JCI)

The higher Chinese hog prices have spurred rising orders of EU pork and could lead to more demand for Chinese pork if the spread widens (global pork prices fall or Chinese prices rise). But unfortunately for U.S. producers the 25% retaliation levied by China against U.S. pork during President Trump's trade war in 2018 remains. That puts them near a $0.20/lb disadvantage to Canada and Europe into China. And high global grain prices have raised global pork prices, making the needed China premiums even higher for Chinese imports to occur.

China's import math is complicated by a sharp "frozen discount" (Chinese prefer chilled/fresh pork over frozen), freight, a 12% regular tariff, and a 9% value-added tax. As stated, U.S. pork also includes a 25% retaliation tariff. And global commodity inflation has driven global pork prices higher as well in Europe, Canada, and the U.S. compared to historical levels. Thus, China's hog price boom is having a muted effect on overseas markets.

4. What Does the Future Hold?

Consider the following watchlist for global pork markets into 2023:

- The recent waves of inflation have peaked. But with the sharp increase in both money supply and debt among key economies, do not expect relative commodity prices to return to the pre-pandemic levels. Recession risk is on globally and could lead to some minor demand pressure in some nations. But pork is relatively resilient to economic recession.
- Feed grain concerns remain a key factor to global pork profitability. The combined disruptions of drought and war drove grain price gains in 2022. Any positive changes to either would have the alternate affect. Global weather forecasts suggest that the 3 year La Niña (warmer equatorial Pacific Ocean) pattern will shift to an El Niño pattern. A warming equatorial Pacific Ocean patterns favors more precipitation across North and South America. This would be broadly positive for grain prices.
- The global drumbeat about climate legislation has increased in recent years. Strict Dutch regulations against nitrogen and ammonia emissions led to months of farmer protests. Many suggest a reduction in Dutch livestock output due to the regulations. More legislation is expected. Canadian Prime Minister Justin Trudeau suggested that Canada adopt a similar rule to that one passed in the Netherlands. The debate of climate change and livestock production will continue. A key question is whether governments will employ punishments (the stick) or rewards (the carrot)

as implements to drive change. Legislation is "the stick" and can lead to unintended consequences by lawmakers naïve to livestock production. The alternative route, "the carrot", would include measures such as carbon markets as means to reduce greenhouse gas reductions. Both are likely to progress in 2023.

- Global demand remains poised to outstrip supply for meat and poultry, including pork. Our model that compares global GDP to global pork demand shows a very high correlation between the two. Global GDP growth fuels demand for pork. Using recent IMF global GDP estimates and comparing to the United Nations Food and Agriculture Organization (UN/FAO) forecasts[2] suggests that over the next 10 years, production will be inadequate to fill global supplies.

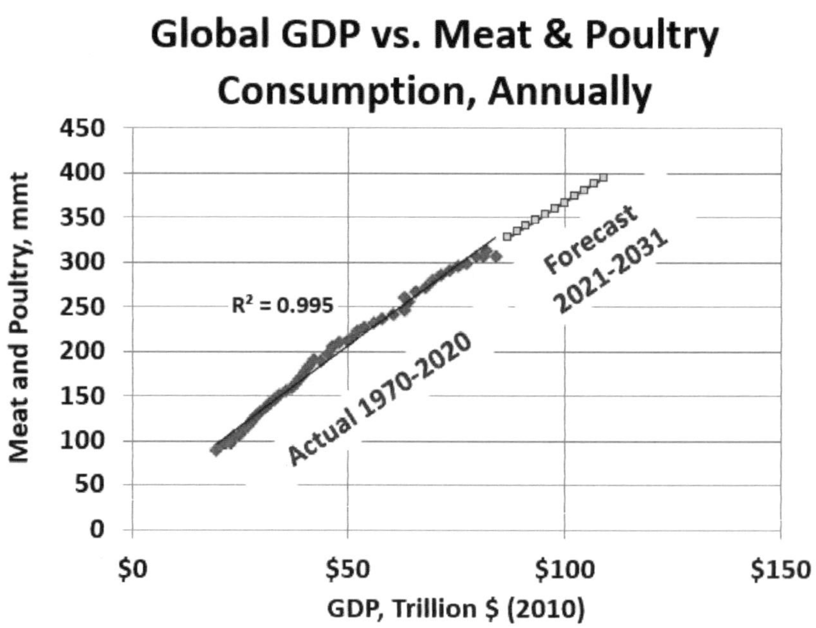

Figure 3. Global GDP vs Meat.Poultry Cons (Source: UN/FAO, IMF, World Bank, AgriTrends)

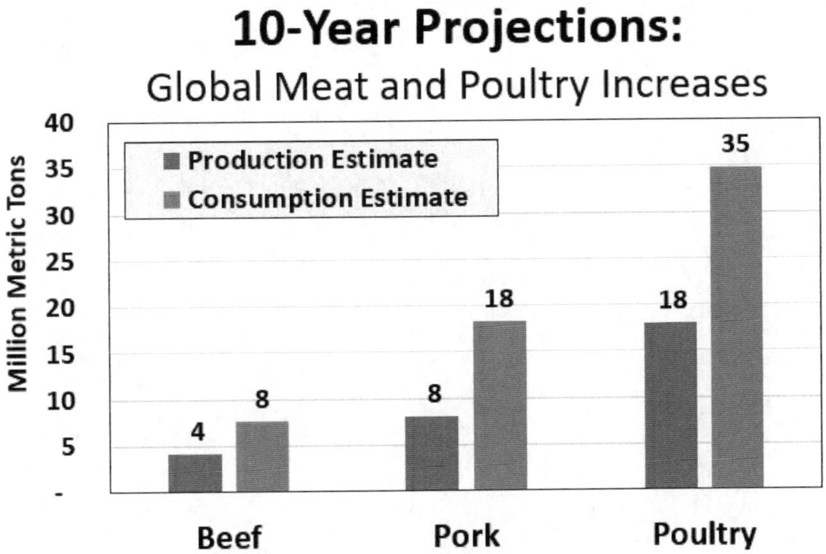

Figure 4. 10 Year Projections, comparison (Source: UN/FAO, IMF, World Bank, AgriTrends)

5. Summary

Global pork supplies remain relatively tight based on disease issues and inflationary pressure on margins. 2023 is the 2nd year in a row to see a smaller global swine herd. This bodes well for global pork producer margins through the year. Uncertainty abounds around global economics, exchange rates, war, weather, and feed grain costs. Overall global supply and demand variables suggest a positive financial year for most global pork producers.

References

1 https://www.wfp.org/news/wfp-chief-warns-grave-dangers-economic-impact-coronavirus-millions-are-pushed-further-hunger

2 https://www.fao.org/3/cb5332en/Meat.pdf

MAPLE LEAF
TM
is delighted to participate in the
2023
Banff Pork Seminar
as a
Premier Plus Sponsor
For information regarding Marketing, Pricing, Settlement, Logistics and Contracting, Call
1-877-937-4647 Brandon (toll-free)
*Produced in Western Canada
Canada's leading producer of premium quality, value-added fresh and further processed meat and meals products.

Maple Leaf Foods Sustainability Program

Mauricio Alanis

Maple Leaf Foods

Overview of Our Sustainability Program

- Maple Leaf's purpose is to Raise the Good in Food. Our vision to be the **Most Sustainable Protein Company on Earth** helps us achieve our purpose.
- Sustainability is not an ESG initiative, but reflects a mindset and a commitment to create shared value to ALL our stakeholders: our consumers, customers, employees, communities, and investors.
- Our sustainability programs and our sustainable meat products are key differentiators and vectors for top line and margin growth for MLF Four pillars support our sustainability efforts: Better Food, Better Care, Better Communities and Better planet.
- **Better Food**: *making real food with real ingredients – eliminating artificial colours and unnecessary antibiotics, reducing sodium levels, and leading in food safety*

  Details:

  i. *Raised without antibiotics* – we are among the largest producers of RWA pork in North America and the largest producer of RWA poultry in Canada

ii. *Product reformulation* – removed ingredients consumers don't recognize (real food movement), focused on pantry ingredients they know and understand, and reduced salt content

iii. *Food safety Leaders* – we are global food safety leaders. MLF has 4 accredited labs and ran over 280,000 food safety tests to detect microorganisms and pathogens in 2021. We adhere to the Global Food Safety Initiative, and had no food safety recalls in 2021.

- **Better Care**: *leading in animal welfare guided by the Five Freedoms of animal care*

  Details:

  i. *Recognized leader in animal welfare* – transitioned 100% of our sows to advanced open sow housing system in 2021. All our pig trailers have hydraulic lifts to reduce the stress of going up and down ramps. All of our hatcheries and processing plants have remote video auditing to monitor and measure animal welfare and to strive for continuous improvement.

  We have installed environmental enrichments in 100% of our Advanced Open Sow Housing sow barns, nursery and growing pig barns

  ii. *Since 2014 we have reduced our antibiotic usage by 96% across our hog operations by optimizing the care we provide to our pigs.*

  iii. *All our animal care programs are guided by the Five Freedoms of animal care* – Freedom from hunger/thirst, from discomfort, from pain, from fear/distress and freedom to express natural behavior *and the Five Domains* – an approach that encompasses the mental state of animals

- **Better Communities**: *delivering shared value, Maple Leaf Centre for Food Security working to reduce food insecurity in Canada, leading in employee safety and diversity and inclusion*

  Details:

  i. *We pursue shared value for all stakeholders* – consumers, customers, suppliers, employees, shareholders

  ii. *Maple Leaf Centre for Food Security* – We launched this Center in 2016 as part of our commitment to reduce food insecurity in Canada by 50% by 2030 and to advocate for critical policies that will create systemic change.

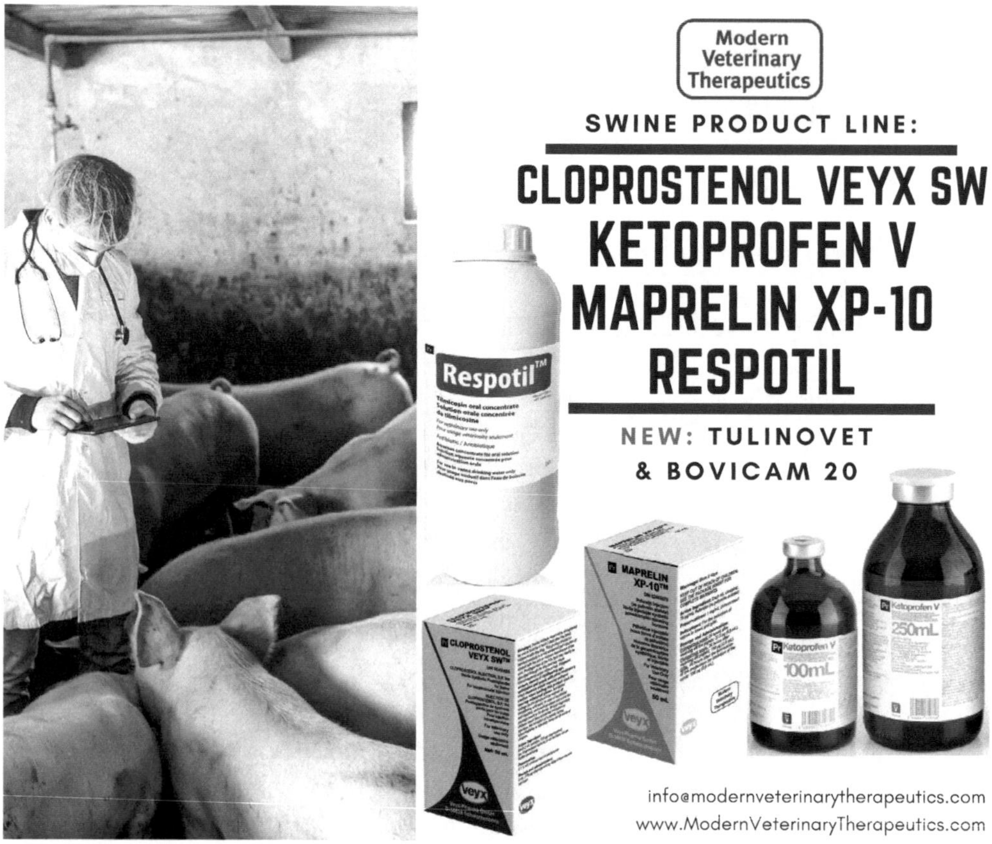

Modern Veterinary Therapeutics
SWINE PRODUCT LINE:
CLOPROSTENOL VEYX SW
KETOPROFEN V
MAPRELIN XP-10
RESPOTIL
NEW: TULINOVET
& BOVICAM 20
Respotil™
Ketoprofen V
100mL
250mL
info@modernveterinarytherapeutics.com
www.ModernVeterinaryTherapeutics.com

BIRTH TO PLATE
Ketchum ID Traceability products have you covered.
Canadian solutions since 1913.
Ketchum MANUFACTURING INC.
SINCE 1913
1245 California Avenue, Brockville
Ontario, Canada K6V 7N5
www.ketchum.ca | 613-342-8455

Circovirus: A Practitioner's Perspective of Present and Future

Ryan Tenbergen DVM, MS

rtenbergen@demetersv.com
Demeter Veterinary Services PC, 5-1249 Seagrave Road, Woodstock, Ontario, N4T 0A8

1. Porcine Circovirus Type 2

Porcine Circoviruses belong to the genus *Circovirus* in the family *Circoviridae* with three serotypes described: PCV1, PCV2 and PCV3. Porcine Circovirus Type 2 (PCV2) is a common pathogen of swine throughout the world and has been shown to be associated with a wide range of clinical presentations and conditions. PCV2 was first identified in the 1990s as an emerging pathogen, and subsequently, caused severe disease in pigs and significant financial losses to swine producers worldwide. The development of vaccines for PCV2 in 2004 and 2006 in Europe and North America, respectively, was a remarkable success story for the global swine industry.

PCV2 is a small single-stranded DNA virus and is, in fact, one of the smallest DNA viruses to infect animals. Although DNA viruses are known to be quite stable, PCV2 exhibits high mutation rates similar to much larger RNA viruses, such as PRRS or Swine Influenza A virus[1]. Circoviruses not only undergo genetic point mutations leading to changes in the virus over time, but have evolved to undergo recombination with other circoviruses. Recombination occurs when two or more circoviruses infect the same cell swapping genetic material and creating the "recombinant" virus. This recombination event between separate viruses is common for circoviruses, with one study finding 33% of 898 sequences from 32 different countries were PCV2 recombinants[2]. PCV2 strains are grouped into several genotypes, with the most common genotypes found in North America being PCV2a, PCV2b and PCV2d, the latter being the most common genotype currently isolated in the United States. However, genotypes a-i have been reported globally in recent years[3]. Figure 1 shows Canadian genotype prevalence data showing the change in genotype prevalence from 2019-2022 as well as prevalence of both single genotype or recombinant genotype infections in pigs.

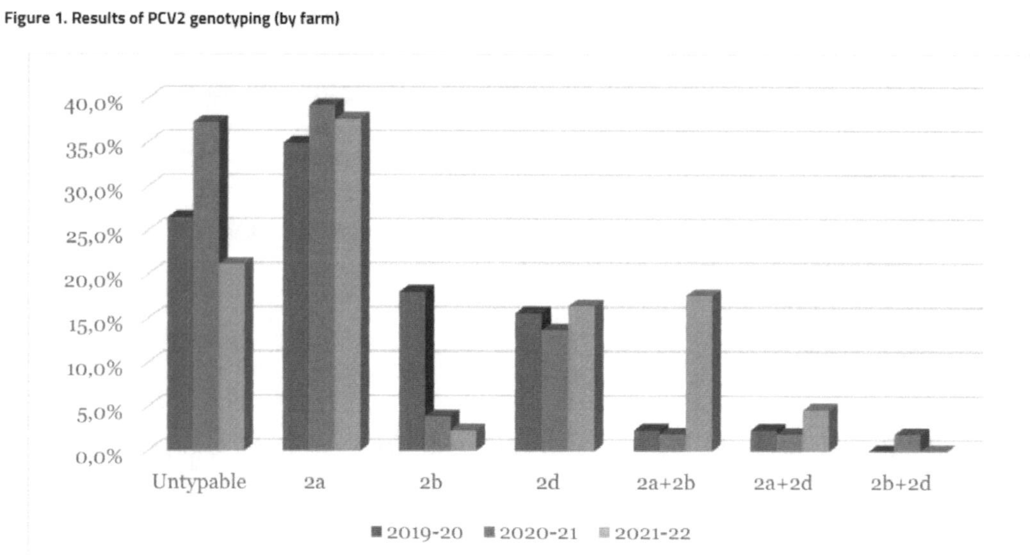

Figure 1. Canadian genotype prevalence data from a diagnostic laboratory (Biovet, 4375 Av. Beaudry, Saint-Hyacinthe, Quebec, Canada).

Although PCV2 vaccines are widely used and, in general, have been effective in preventing and controlling PCV2-associated diseases, there are reports of clinical disease presentations in vaccinated populations in recent years[4, 5]. In theory, minor changes in the capsid protein, the virus part recognized by the immune system, may allow some of the new PCV2 recombinant viruses to escape the pigs' immune system after vaccination leaving it susceptible to clinical disease. However, to date, PCV2 vaccine, both original and new, have proven effective against circulating genotypes in the field[3], with suspected vaccine failures being related to other factors besides vaccine efficacy.

Monitoring strategies for PCV2 include processing fluids, PCRs on tissues or blood and serology on blood. Being a ubiquitous virus in swine populations, the significance of PCR-positive processing fluids and effects on performance in swine herds is being investigated, but it remains a simple and quick way to monitor circovirus in swine herds. The presence of viremia in growing pig populations and its significance has been debated, but multiple studies has shown the negative growth performance effects of circovirus viremia in growing pigs (Figure 2)[6].

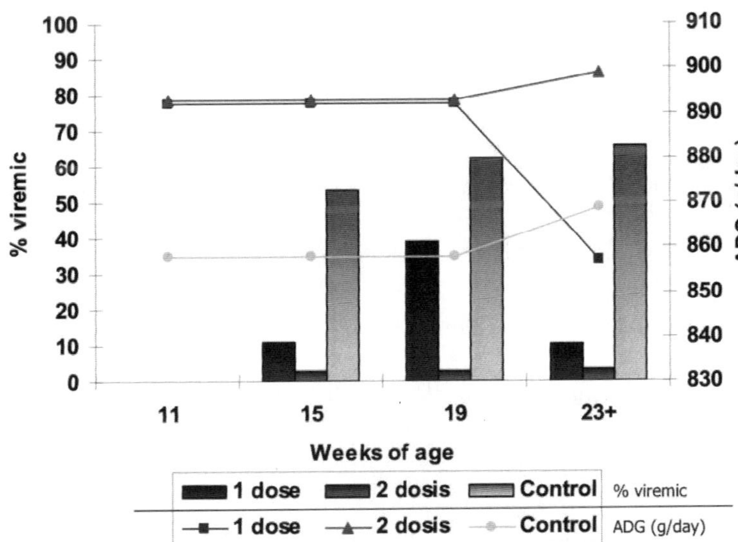

Figure 2. Viremia and ADG in the grower-finisher phase comparing 1-dose versus 2-dose PCV2 vaccination programs (Reindl M, 2010).

2. Porcine Circovirus Type 3

PCV3 was first reported in the literature as being potentially pathogenic in swine in 2016[7, 8]. Recombination of circoviruses, potentially mammalian and avian, has been suggested for the emergence of this newly described circovirus in pigs[9]. In fact, the capsid and replicase proteins of PCV3, those responsible for recognition and response by the immune system, share only 26-37% and 48% similarity with PCV2, respectively[7]. In addition, PCV3 genome analyses reveal low genomic variability among viruses suggesting a generally recent origin[9]. In sharp contrast to the history of PCV2, whose causal association with significant clinical disease was debated for a long time, PCV3 was rapidly accepted as a potential swine pathogen worldwide by the veterinary community showing several parallelisms between the two circoviruses. Through testing of historical laboratory samples, PCV3 is thought to be an endemic virus, like PCV2, and has most likely been circulating in pig populations for years[8].

3. Conclusion

In order to stay ahead of circovirus-associated diseases and avoid the potential of significant losses associated with clinical disease, we need to rethink circovirus monitoring and vaccination strategies against it. Although current commercial vaccines provide good cross-protection against circulating PCV2 genotypes, instances such as continued PCV2 recombination events, or the newly described and genetically distinct PCV3 serotype, will continue to challenge swine populations and vaccines. Simply because of the enormous historical economic impact that PCV2 has had in swine populations worldwide, PCV3 deserves and is gathering attention and warrants further investigation into its significance and role in clinical disease. From a practitioner's perspective, do not forget about circovirus and invest in circovirus monitoring on your farm to stay ahead of the virus! If you suspect circovirus is contributing to health challenges on your farm, you need to work with your herd veterinarian to confirm if it is truly playing a major role on your farm.

References

1. Ssemadaali MA et al. Genetic diversity of porcine circovirus type 2 and implications for detection and control. Research in Veterinary Science 2015;103:179.
2. Franzo G et al. Phylodynamic analysis of porcine circovirus type 2 reveals global waves of emerging genetoypes and the circulation of recombinant forms. Mol Phylogenet Evol. 2016;100:269.
3. Franzo G and Segales J. Porcine Circovirus 2 Genotypes, Immunity and Vaccines: Multiple Genotypes but One Single Serotype. Pathogens 2020;9:1049.
4. Xiao CT et al. Complete Genome Sequence of a Novel Porcine Circovirus type 2b Variant Present in Cases of Vaccine Failures in the United States. J Virol. 2012;86(22):12469.
5. Johnson C. PCV2d outbreaks and control. AASV Proceedings 2021;9-10.
6. Reindl M (2010). A comparison of growth performance and viral shedding in pigs vaccinated with porcine circovirus type 2 vaccine. [Unpublished Master of Science thesis]. University of Guelph.
7. Phan TG et al. Detection of a novel circovirus PCV3 in pigs with cardiac and multi-systemic inflammation. Virology Journal 2016;13:184.
8. Palinski R et al. A Novel Porcine Circovirus Distantly Related to Known Circoviruses Is Associated with Porcine Dermatitis and Nephropathy Syndrome and Reproductive Failure. J Virol. 2016;91(1)
9. Franzo et al. The analysis of gene composition and codon bias reveals distinctive patterns between avian and mammalian circoviruses which suggest a potential recombinant origin for Porcine Circovirus 3. PLoS ONE 2018;13(6):e0199950.

Connect with us!!

www.slt.ca

204.326.6380

Steve's Livestock Transport

steveslivestock

STEVE'S
LIVESTOCK TRANSPORT

Streptococcus suis and *Escherichia coli* in the Nursery: A Health by Nutrition Approach

Brooke N. Smith, DVM, PhD

Cargill
brooke_smith1@cargill.com

1. Introduction

In current swine production systems, one of the most significant stressors faced during the nursery transition is disease. While not a new challenge, increased virulence of today's pathogens affecting our swine populations make them more contagious, more clinically severe, and more difficult to address with traditional treatments or management practices. For that reason, alternative strategies such as nutritional intervention are being investigated and employed to support pigs during a disease challenge. Two bacterial pathogens that consistently pose an issue for nursery pigs in the US and Canada are *Streptococcus suis* and *Escherichia coli*, both contributing to significant financial losses due to increased mortality and increased treatment costs. The discussion below describes the characteristics and clinical manifestations of these two pathogens and how they are currently treated and managed; with that information in mind, potential nutritional interventions and strategies can be identified as mitigants for the clinical disease these pathogens cause.

2. Streptococcus suis

2.1 Disease Characteristics, Clinical Signs, and Diagnosis

Streptococcus suis (*Strep. suis*) is Gram-positive, alpha-hemolytic bacterium and is among one of the leading causes of death in post-weaned pigs 5-10 weeks of age. In North America (NA), there are several residential serotypes with serotypes 1, 2, and 1/2 being the most predominant[1,2]. It is important to note that most of the NA serotypes are centralized to the Midwest United States and distribution and virulence of these serotypes differ between the US and Canada[1].

Streptococcus suis can affect any age of pig, but most commonly results in disease in the nursery phase. It is transmitted both vertically during parturition and horizontally by infected individuals. *Streptococcus suis* is considered a facultative pathogen, so a main source of exposure are healthy, asymptomatic carriers. Establishment of clinical disease depends on several factors including concurrent viral disease, crowding, ventilation, weaning, and diet changes. Palatine tonsillar epithelium is the most common site of colonization and dissemination for *Strep. suis*, though there is some evidence of certain field serotypes (e.g., serotypes 9 and 2) being able to colonize the intestinal tract[2]. While morbidity and mortality have been historically lower for *Strep. suis* as a single agent pathogen, the virulence of current field strains as well as the likelihood for concurrent viral or bacterial infections increases the severity of typical outbreaks today.

Clinical signs and gross lesions caused by *Strep. suis* are dependent on the body systems involved and the current state of infection. Early signs are those of generalized illness and those associated with septicemia. Acute deaths without premonitory signs may occur as well. Common presentations for *Strep. suis* in NA include polyarthritis and meningitis. *Streptococcus suis* is typically diagnosed presumptively by case history and observation of clinical signs, though bacterial culture is required for definitive diagnosis. Preferred samples are those from acutely ill animals and are dependent on the body systems involved. Samples from tonsils or nasal cavities are not considered diagnostic since nearly 100% of commercial pigs are carriers of at least one serotype of *Strep. suis*[1].

2.2 Traditional Control Strategies and Current Status

Prompt recognition of clinical signs and immediate parenteral (i.e., injectable) treatment with antibiotics is often identified as the best approach to maximize survival. However, when faced with many at-risk animals or in flows of pigs with a previously diagnosed *Strep. suis* problem, water-delivered antibiotics are often utilized. Whenever possible, microbial sensitivity tests should be employed on any *Strep. suis* that is cultured from diagnostic samples to aid in antibiotic selection. In addition to antimicrobials, anti-inflammatory medications such as dexamethasone or aspirin are utilized as adjunct therapies due to the severity of inflammation *Strep. suis* infections can cause. Even with proper treatment, most animals that progress to severe septicemia or neurological disease will perish. Regarding today's field serotypes, while there doesn't seem to be significant levels of antimicrobial resistance, their increasing virulence often results in the requirement of multiple rounds of antibiotic therapy to get nursery pigs through an outbreak while still incurring high levels of morbidity and mortality. Autogenous bacterins can be used as a preventative measure in the field but are relatively ineffective in preventing outbreaks[1].

3. Escherichia coli

3.1 Disease Characteristics, Clinical Signs, and Diagnosis

Escherichia coli (*E. coli)* is a Gram-negative, flagellated bacilli for which both commensal and pathogenic strains exist. In the nursery, *E. coli* presents clinically either as diarrhea or enterotoxemia (otherwise known as edema disease or ED), the latter of which results in higher mortality rates. *Escherichia coli* are classified by a number of schemes including serotype, pathotype, pilus type, and toxin type. In the US and Canada, the most commonly diagnosed pathogenic field strains of *E. coli* are K88/F4 and F18. While K88/F4 has previously been more frequently diagnosed in nursery pigs, and still is in Canada, F18 outbreaks are increasing in both frequency and severity within the US. Pigs most often become infected via ingestion of *E. coli* shed from immune carrier dams or from their environment. Pathogenic *E. coli* bind to specific receptors on the microvilli of enterocytes in the small intestine where they colonize, proliferate, and produce various enterotoxins that cause clinicals signs; the expression of theses receptors is influenced by both genetic and age factors.

Clinical signs are associated with the type of enterotoxins produced, but in general pathogenic *E. coli* cause a toxin-mediated hypersecretory diarrhea resulting in severe dehydration and metabolic acidosis, which may progress to terminal sepsis and endotoxemia. Several external factors can influence the severity of infection and clinical signs of *E. coli* infections as well, including a cool and/or damp environment, reduced feed intake, transitions to plant-based diets, commingling or double stocking, and poor environmental sanitation. Edema disease caused by enterotoxemia typically occurs 1-2 weeks post-weaning and is often characterized by acute deaths of previously healthy pigs or anorexia, neurologic signs (e.g., ataxia, stupor, paddling), and soft tissue swelling, most noticeable in the face and eyelids. Diarrhea is not usually present in pigs with typical ED. Diagnosis of *E. coli* is based on observation of typical clinical signs and isolation of pathogenic *E. coli* on bacterial culture from small intestine samples. Since *E. coli* is a normal part of the small intestine microbial flora, histological evidence of adherent *E. coli* on the brush border of the small intestines can be useful in reaching a final diagnosis. For neurologic presentations of ED, other common causes of neurologic disease in pigs should be ruled out including *Streptococcus suis, Glaesserella parasuis,* water depravation, or acute toxicities[4].

3.2 Traditional Control Strategies and Current Status

Diarrhea caused by pathogenic *E. coli* will typically persist until appropriate medical intervention is implemented. Unfortunately, there is a high degree of antimicrobial resistance in today's field strains of pathogenic *E. coli*, particularly for F18, making traditional antibiotic choices not as effective[4,5]. For that reason, basing antimicrobial selection on sensitivity data collected from diagnostic samples is recommended. Addressing external factors such as environment or husbandry should also be considered an integral part of *E. coli* treatment on farm. Similar to *Strep. suis*, *E. coli* bacterins are available as well as live attenuated or wild-type avirulent *E. coli* vaccines for prevention, but their success is variable.

4. Nutritional Approach to Disease: Considerations for Inflammation and Pathogen Characteristics

With the understanding of how these two bacterial pathogens cause disease in nursery pigs, attempting to control their clinical manifestation through nutrition is based on several factors. First, it depends on the type of physiological response they cause as activation of pro-inflammatory pathways and pro-oxidative pathways are not mutually exclusive, though will have overlapping activators and functions. Oxidative stress occurs when there is an imbalance between the production of reactive oxygen species (ROS) and their elimination via protective mechanisms, either by disruption of protective cell pathways or by increased production. These ROS are generated under numerous conditions including normal metabolism, active immune responses, and environmental stressors (e.g., heat stress). On the other hand, inflammation is a protective response to injury, infection, or irritation and is driven by the immune system. It is comprised of several physiologic responses including pro-inflammatory protein production, leukocyte infiltration into host tissues, vasodilatory effects, activation of clotting factors, and generation of ROS. Both processes are essential for clearance of pathogens, but in doing so produce some degree of host tissue damage. So, while related through shared inciting agents and cellular responses, oxidative stress and inflammation can occur independently and thus may not be responsive to the same nutritional interventions depending on their mechanism of action.

Another factor to consider for nutritional approaches to disease are how dynamics of infection for individual pathogens can make them more or less sensitive to direct intervention via feed. For example, enteric pathogens like *E. coli* have direct contact with dietary components within the intestine, making neutralization or inhibition more likely. Alternatively, *Strep. suis*, typically colonizes other non-enteric mucosal sites and thus has less contact time with dietary components, making nutritional interventions that modulate the host response to disease on a systemic level a more viable approach.

4.1 Nutritional Technologies – Pathogen vs. Host Response Targets

Nutritional technologies that target pathogens directly are those ingredients and/or nutrients that either directly cause damage to the pathogen, interfere with its ability to bind receptors or colonize target host tissue sites, or alter the environment in a way that does not support growth or dissemination of the pathogen. Two examples of commonly used nutritional additives or technologies that can result in direct damage to pathogens by altering bacterial cell membranes include essential oils (EO) and antimicrobial peptides (AMP)[6]. Some technologies that interfere with the pathogen's ability bind receptors or colonize include zinc oxide and mucilages. These compounds act on the intestinal mucosa directly by reinforcing intestinal epithelial integrity or through physical blocking of binding receptors. Dietary fiber, particularly structural carbohydrate sources, should also be considered as a protective nutritional strategy as it promotes these same effects at the level of the intestinal mucosa while also providing substrate to commensal microbes that can act via competitive exclusion on pathogenic bacteria. Another way colonization of pathogens can be disrupted is through a process called quorum sensing inhibition, which is the prevention of bacteria-to-bacteria communication within a microbial population by ingested bioactive compounds resulting in decreased expression of virulence factors and decreased biofilm and fimbriae production[5].

Nutritional technologies that target the host immune response are largely comprised of compounds, either plant- or animal-derived, that exert some level of biological activity within the host that is not confined to the gastrointestinal tract. For plant-derived (i.e., phytogenic) compounds, these include phenolic compounds, carotenoids, steroids, thiol compounds, tocopherols (i.e., Vitamin E) and ascorbate (i.e., Vitamin C), among others. These compounds exert both anti-inflammatory and anti-oxidative properties, though some on this list are utilized more than others in livestock nutrition formulations. Among primarily anti-oxidant ingredients, Vitamin E plays an important role in ROS scavenging and is often supplemented at higher than NRC recommended levels (≥200 IU/kg) in the diets of disease-challenged pigs. Among ingredients that express primary anti-inflammatory properties, phenolic compounds represent a diverse source of biological activities including immune modulation, direct antimicrobial/antifungal activity, and promoting gastrointestinal mucosal integrity. One particular pro-inflammatory pathway that some phenolic compounds are able to modulate is the transcription and expression of nuclear factor kappa B (NFκB) and nuclear factor erythroid 2-related factor (Nrf-2). These pathways work in opposing ways in response to inflammation and oxidative stress. NFκB induces synthesis of pro-inflammatory proteins like cytokines and chemokines, adhesions molecules, and enzymes which will promote inflammation including those that generate ROS while Nrf-2 acts as a positive regulator for expression of anti-oxidant and detoxification enzymes[5]. When NFκB expression is reduced and Nrf-2 expression is increased, there is a reduction in the severity of the inflammation induced by the inciting agent, thus reducing host tissue damage, and re-establishment of cellular redox homeostasis.

5. Conclusions

Disease in the nursery continues to reduce survivability and productivity during early stages of post-weaning growth. Changes in pathogen virulence and susceptibility to traditional medical interventions means alternative support strategies such as nutritional intervention can improve clinical outcomes. The pathogen causing disease and the type of physiologic response it causes should be considered when choosing nutritional technologies or feeding strategies.

References

1. Gottschalk, M., Lacouture, S. (2015) Canada: Distribution of *Streptococcus suis* (from 2012 to 2014) and *Actinobacillus pleuropneumoniae* (from 2011 to 2014) serotypes isolated from diseased pigs. Can Vet J. 56, 10, pp. 1093-1094.
2. Segura, M., Aragon, V., Brockmeier, S. L., Gebhart, C., de Greeff, A., Kerdsin, A., O'Dea, M. A., Okura, M., Salery, M., Schultz, C., Valentin-Weigard, P., Weinert, L. A.,

Wells, J. M., Gottschalk, M. (2020) Update on *Streptococcus suis* Research and Prevention in the Era of Antimicrobial Resistance: 4th International Workshop on *S. suis* [Conference Report]. Pathogens, 9, 374: pp.1-39.

3. Fairbrother, J.M, Nadeu, E (2019) Colibacillosis. *In:* Diseases of Swine – Eleventh Edition. Eds. J.J. Zimmerman, L.A. Karriker., A. Ramirez, K.J. Schwartz. G.W. Stevenson, J. Zhang. John Wiley & Sons, Inc., Hoboken, NJ, pp. 807-832.
4. Lunha, K., Leangapichart, T., Jiwakanon, J., Angkititrakul, S., Sunde, M., Jarhult, J.D., Strom Hallenberg, G., Hickman, R.A., Van Boeckel, T., Magnusson, U. (2020) Antimicrobial Resistance in Fecal *Escherichia coli* from Humans and Pigs at Farm at Different levels of Intensification. Antibiotics 9, 662, pp. 1-11.
5. Yang, C., Kabir Chowdhury, M.A., Hou, Y., Gong, J. (2015) Phytogenic compounds as alternatives to in-feed antibiotics: potentials and challenges in application. Pathogens, 4, 1, pp. 137-156
6. Liu, H., Cao, X., Wang, H., Zhao, J., Wang, X. Wang, Y. (2019) Antimicrobial peptide KR-32 Alleviated *Escherichia coli* K88-induced Fatty Acid Malabsorption by Improving Expression of Fatty Acid Transporter Protein 4 (FATP4). J. Anim. Sci., 97, 6, pp. 2342-2356.

OLYMEL
Feeding the world
Olymel S.E.C., a world class Canadian leader in the processing of Pork and Poultry, with more than 11,500 employees throughout Canada, is proud to have a state of the art operation in Red Deer, Alberta. We pride ourselves in supporting the agricultural economy, and are equally proud to sponsor the Banff Pork Seminar.
Partnering with our pork producers is a priority and together, we offer the world quality products with superior taste. Olymel has a single motivation and mission:
Together. we feed the world.
olymel.ca
7550 40th Avenue Red Deer, AB T4N 6R7 | 403 343-8700

800-500-7978
info.canada@minitube.com
Leading assisted reproduction technology into the future
Innovative Minitube products professionalize and optimize the processes in stations and laboratories worldwide. Our knowledge can lead you to increased efficiencies. Ask us how!
minitube
www.minitube.com

fcc
Partner with the only lender 100% invested in Canadian agriculture and food.
DREAM. GROW. THRIVE. | fcc.ca

Feeding the Sow & Managing the Farrowing Environment to Improve Nursery Pig Performance

Casey Neill, Beau Peterson, Jorge Estrada and Andrea Hanson

Carthage Veterinary Services

Having world class wean to finish performance starts with the quality of wean pig the producer receives from the sow farm. Work done by Main et al., 2004, demonstrated that wean age will have major impacts on mortality, average daily gain and profitability in the wean to finish barn. An older wean pig also means that the weaning weight will also be heavier which will help the wean pig start on feed and ultimately thrive. Achieving an older, heavier wean pig can occur with intense focus on management and care of the sow during lactation. Some of the areas that we will discuss to achieve this goal is sow lactation feed intake, sow body condition management, water intake, lactation nutrition and farrowing house management.

Sow lactation feed intake is critical to drive milk production and successful conception of the next litter. There are many factors that will drive sow lactation feed intake including sow body condition. Work done by Young et al., 2004, and Bryan et al., 2013 demonstrated that over-conditioned sows will have a lower feed intake compared to sows in a normal body condition. Reduced feed intake drives reduced milk production which negatively affects wean pig weights.

To get the neonatal pigs off to a great start they need a minimum of 250 ml of colostrum (Quesnel et al., 2012). In order to achieve sufficient colostrum production, the sows need to have access to fresh water. Water is an impactful area often overlooked in the farrowing house. Sow water intake during lactation should be a minimum of 25 liters per day at peak lactation. It takes 2.5 liters of water to producer 1 liter of milk. Therefore, a sow must consume 35 liters of water per day to support peak milk production of 14 liters per day.

Lactation diet formulation can also drive milk production and weaning weight. Formulating to meet the amino acid levels and match feed intake is non-negotiable. Sow research conducted at Carthage Veterinary Service with Dr. Laura Greiner, PIC, University of Missouri and Ajinomoto on a 6,000 head sow research farm in Illinois, was the first of its kind to have large scale sow research to evaluate the effects of dietary lysine levels on weaning weight. The researchers demonstrated that as SID Lysine intake increased to 60 grams/day, weaning weights improved while sows lost the least amount of body weight and had the shortest wean to estrus interval. Producers must watch closely their average lactation feed intake as gilts will consume 10% to 15% less than older parity sows. Consequently, knowing the lactation feed intake of a herd is critical when designing a feed program. If only the average is considered, you will be under feeding the younger sows and gilts. Some producers have decided to feed to the gilt requirement as they are the future of the herd which means the percent of SID Lysine in the lactation diet will be higher. It is recommended to feed a gilt lactation diet for new start up farms. With a higher SID Lysine diet the inclusion of protein sources like soybean meal should be formulated carefully. Sow research with high levels of soybean meal (30%) showed reduced lactation feed intake (Gourley et at., 2020). Farrowing house temperature also affects lactation feed intake. It is critical to keep the room temperature comfortable for the sow to promote ad libitum feed intake but not chill the newborn pigs. The last thing we would want to do is cause a draft on the baby pigs as this could cause them to scour. Starting the room at 74 degrees and gradually lowering the temperature generally keeps sows comfortable and encourages suckling pigs to stay warm under the heat lamp or heat mat.

To have great nursery pig performance it all starts at the sow farm. Having the best quality wean pig producers must focus on proper sow body condition, feed and water intake, amino acid intake, formulation, and farrowing house temperature.

What to Expect from an AI Dose & Best Practices on Farm

Kyla Ripley

Carlo Genetics Inc

Production rates and consumption demands for pig meat are continuously increasing worldwide. Production must, therefore, be sustainable and highly efficient while done in a livestock compatible environment to avoid losing competitiveness[1]. In this context, increasing genetic progress of the swine population to produce more and better meat with fewer bred animals is critical to diminish environmental pressure, both in terms of feedstuff production and manure handling[2].

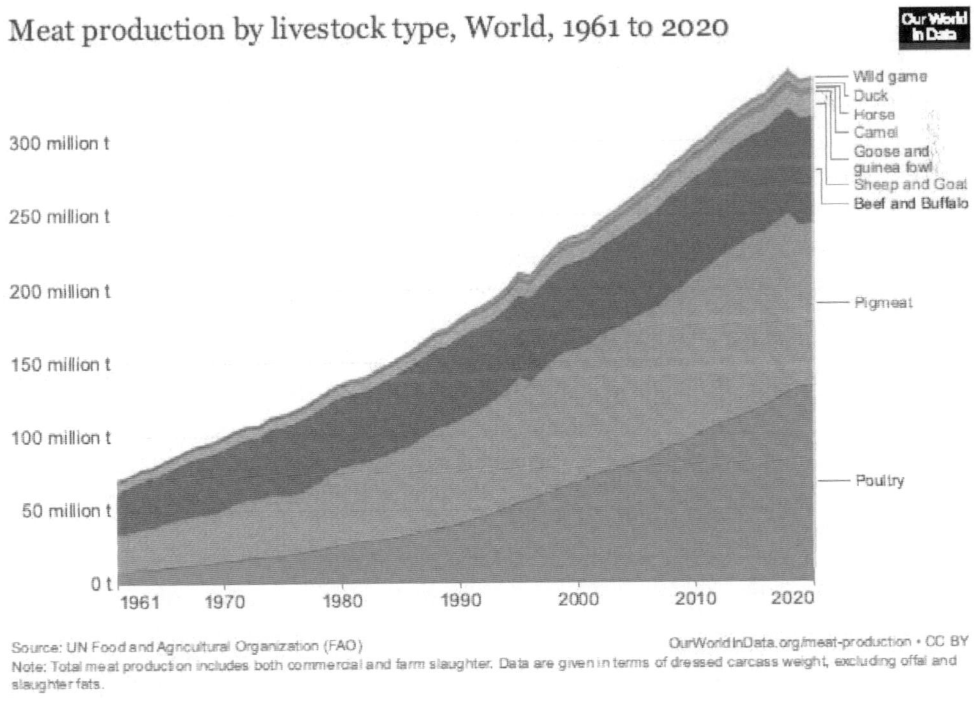

Figure 1. Meat production by livestock type, World, 1961 to 2020

Consequently, the current picture of AI-centers is moving from classical small centers housing a few boars usually integrated within breeding farms, towards large, independent and technically highly specialized centers housing a large number of healthy and genetically superior boars that produce large daily numbers of ready to use high quality liquid stored AI doses following strict regulations and guidelines to prevent disease spreading[3]. These professional AI Centers have moved to more accurate and precise semen assessment by incorporating technologies such as CASA (computer assisted semen assessment) using 4 and 5K camera technology and sophisticated phase contrast microscopy incorporating such items as chambers slides which allow free movement of sperm cells and more correct evaluation of motility and morphology at line production speed. This along with such technology for concentration verification and membrane integrity assessment as the Nucleocounter SP100 from Chemometic, allow for precision dose creation in required formats thus ensuring the more efficient use of desired high indexing sires in a breeding program and ultimately assisting in driving that progress out to commercial production. When artificial insemination in swine started in practice some 30+ years ago, the driver was to aid in the prevention of venereal disease while delivering as good a result as natural mating which was by and large the norm. Its use had the added benefit of not having to manage boars on farm which could be challenging.

In the last decade the global adoption of artificial insemination has led to the reality that in Canada some 95% of matings are by artificial insemination with approximately 30% of those estimated to be by PCAI (post-cervical-AI).

Table 1. AI rates and technique in the top 10 pork-producing countries[4].

	Pork production 2018 (in 1,000 metric tons)*	Estimated portion AI (of total breeding)	Estimated portion PCAI (of total AI)	Source
China	54,650	40%**	50%	Minitube
EU	24,050	95%	0–60%***	Various
USA	12,166	90%	30%	Minitube
Brazil	3,675	95%	60%	F. Bortolozzo
Russia	3,050	85%**	25%	Minitube
Vietnam	2,800	90%	30%	D.H. Bui, X. Le Thi
Canada	2,015	>95%**	30%	Minitube
Philippines	1,600	85**	5%	PIC
South Korea	1,321	90%	30%	PIC
Mexico	1,305	90%	25%	Minitube

PCAI: post cervical artificial insemination.

* https://www.statista.com/statistics/273232/net-pork-production-worldwide-by-country/.

**>90% in professional farms.

***0%: Denmark, Germany, The Netherland.

The Role of an AI Centre in Your Breeding Program

The AI Center carries the role of the production of cost efficient, safe semen doses with high fertilizing capacity to increase the use of boars of high genetic value **(5)** in cooperation with genetic companies who produce the boars. AI Centers in Canada are under the governance of federal control monitors through CFIA to ensure the health of boars entering stud and therefore conform to strict biosecurity measures to ensure ongoing security of the supply chain.

Producers on sow breeding farms should expect that the AI Center they work with is operational within a robust Quality Assurance program. A typical and standard QA program will include operating procedures detailing set limits and monitoring of:

Motility: Fresh ejaculates, End doses through storage, End doses at listed expiration date

Morphology: Boar evaluation by fixed or stained evaluation, Fresh ejaculates if using CASA auto morphology, End dose random sampling by fixed or stained evaluation

Concentration: evaluation of fresh ejaculates and assurance of final extended doses

Bacteriology Sampling routine: environment, purified water, extender, end doses

Hygiene: Detailed program designed to limit as much as possible the presence of bacteria.

Biosecurity Program: Includes requirements to enter boars but extends to working with veterinarians to ensure bio-secure delivery of doses in so far as these are within their control. E.g., Local routes with organized drop sites away from direct farm deliveries. It is important to work with you AI Center to establish this as a routine.

Globally, many countries, groups and breeding organizations have established minimum standards in practice for the production of AI doses[4].

Table 2. Minimum standards for the use of liquid preserved boar semen in 11 pig breeding organization worldwide.

Semen Standards	A	B	C	D	E	F	G	H	I	J	K
Total motility (%)											
Fresh	70	80	75	70					70		80
Expire day	60	70		65			60		50	60	70
Progressive motility (%)											
Fresh		80	70			70					
Expire day		70			70	60	50	70		45	
Morphological abnormal sperm (%)	30	20	25	25	25	15	30	25	30	30	
Cytoplasmic droplets (%)		20		15		15	20		30		
Agglutination (%)	20			30		15	30	30			
Sperm/dose ($x10^9$)											
Cervical insemination	1.8	2.0	2.3	1.8	4	3		3	2.2	1.3*	1.5*
Post cervical insemination		1.5	1.4		2.2	1.8		1.5	1.1		
Storage duration (days)	3–7	6	2–5	3	2–3	7		5–6	5	4	4
Extender	s, l	l	l	s		l	l	l	l	s	s
Bacterial contamination (CFU/mL)	300	0	0		1000		0	0	30		300

Communication

Good communication with your AI Center helps you understand the product you are buying so you are able to make best use of it on your farm. A conversation with your AI supplier could include questions such as these to help you understand their work routine and the product you will receive:

a) What is the final validation date of doses? How often is motility checked in storage and will I be called if there is a finding below the cut-off limit?

b) What is the cut-off limit for good morphology of the boars I will receive?

c) What type of packaging is used to deliver doses to my farm? Does it protect from sunlight and how is the temperature controlled in delivery?

d) Does the routine veterinary monitoring of the Ai Centre include testing of the boars? What tests and how often? Your veterinarian could speak with the AI Centre veterinarian on your behalf as well on this question.

e) Is the extender you use a short term or long-term extender? Short term extenders are typically 3-4 days and long term go out to 7 days. Understanding this helps you organize your orders as you need.

f) Do you monitor doses for bacterial contamination?

g) Do you participate in external verification of quality control points?

My AI has Been Scheduled for Delivery – Now What?

Secure delivery is the goal for AI dose delivery to the farm. Communication with your AI Centre before first delivery should confirm drop location and expected time of arrival.

Temperature control of AI doses transfers from the AI center driver to you at the moment of drop off so it is important to be prepared. Modern extenders used in production of doses contain buffers to protect cells against the effects of fluctuations in temperature and pH. As opposed to early days of AI, doses can safely be transported in a wider temperature band as a result 14 – 19 °C is very acceptable but temperatures outside of this no longer mean the dose is not usable but rather may need to be used in a shorter timeframe. Making a call to your AI supplier to discuss is the best practice. They will know best what the limitations of the extender they are working with are and be able to discuss best options.

Your drop off should ideally be equipped with a semen storage unit that is placed in an insulated room/building that is heated in winter. Just like a standard refrigerator, a semen storage unit will not safely store your doses if it is in a building or shed at -25°C on a February day in Canada!

The very definition of "drop-off" should be as a temporary point of storage. Working with your AI Centre to understand delivery time will ensure you can get your doses from your drop-off and safely in to your farm soon after delivery.

Your semen storage unit on farm should be located in a clean area like the office that is away from the dust of the farm. Semen storage units should have a circulation fan and wire racks to allow air flow when doses are unpacked. Keeping only AI doses in your AI storage unit is best practice as it helps to preserve the cleanliness and functionality of the unit. Temperature control at 17°C is the standard recommendation for fresh liquid doses.

Knowing how the unit is functioning through storage helps you monitor any possible trouble spots that could affect the outcomes of breeding.

Best practices should include:

a) A simple max/min thermometer with a scheduled routine to check temperature two times per day can save lost doses and lost litters. More sophisticated systems offer a temperature probe and alarm system to call out if there are temperatures detected out of range for quality storage.

b) Organize doses on shelves taking care to lay them out to maximize the air flow functionality of the unit for most even temperature control.

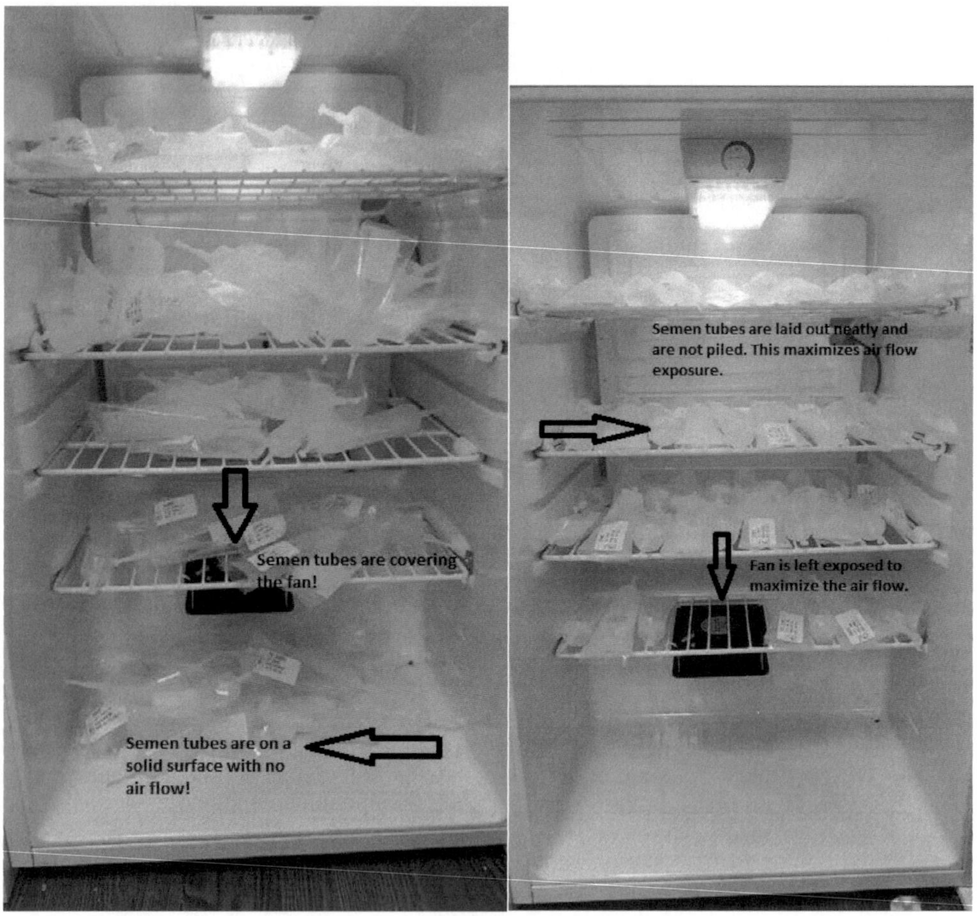

Figure 2. Improper storage of semen dose on left & ideal storage on right.

c) Know the capacity of your unit. If you are a batch breeder ordering 600 doses in a delivery, a small unit with max capacity of 200 doses is not going to serve your needs even if for short term.

The ROI of an added storage unit to a 2000 sow farm even in modest markets can represent some $5000 (As per personal communication with the late Dr. Hanneke Feitsma, DVM; March 6, 2017 and sharing of calculator).

Table 3. Calculator ROI added semen storage cabinet; Dr. Hanneke Feitsma as presented at Manitoba Swine Seminar

# Storage cabinets ($ 876/cabinet)	1	2	Difference
# Sows	2,000	2,000	
FR	87.0%	87.5%	0.50%
TNB	14.00	14.10	0.1
Piglets/year	24,360	24,675	
Doses/cycle	2	2	
Cycle/sow	3	3	
No. Doses/year	12,000	12,000	
Cost per insemination	$ 8.00	$ 8.00	
Storage costs	$ 876.00	$ 1,752.00	
Insemination costs (dose + storage)	$ 96,876	$ 97,752	$ 876
Income per pig	$ 18.00	$ 18.00	
Income from pigs (sold $100/pig)	$ 438,480	$ 444,150	$ 5,670
ROI			**$ 4,794**

While in storage and as you prepare to use the doses on farm, take only doses needed to the breeding barn. Use of a small picnic cooler will serve to keep doses clean and temperature secure. While extenders do protect semen cells, temperature fluctuations with warming and cooling can cause damage to cells leading to less-than-optimal fertility outcomes. Therefore, best practice is careful management of farm storage and handling to avoid these.

To Turn or Not to Turn

Old thinking for dose storage made the recommendation to turn or even rock and mix doses in storage with the belief that the cells needed to be actively exposed to the extender and the food it provided the cells. It has long been assumed that sedimentation of liquid preserved spermatozoa during storage is detrimental, as toxic metabolites accumulate and a local pH shift takes place[5]. In order to prevent this, it was then recommended that sow farms adopt a routine to turn doses and some tools were even developed to mechanically rotate doses on a timer.

However, in contrast to this old practice, more recent studies in 2015 have shown that turning of doses in storage in fact has a negative impact on quality[6]. The findings indicated a loss of membrane integrity at day 3, and loss of motility and changes in sperm kinematics at day 5 with sperm motility and velocity also impacted in continuously rotated samples. It is concluded that doses in storage on farm are better left undisturbed until used.

Conclusion

Quality Assurance programs and embracing advancing technologies has led to extensive quality control in modern AI Centers globally. Good communication between producers and AI suppliers can help ensure the best outcomes on farm.

Though fertility cannot be guaranteed, simple best practices of handling and storage can help in achieving the best outcomes possible.

References

[1] **FAO. World Agriculture**: towards 2015/2030. A fao perspective. Edited by Jelle Bruinsma. 2003. ISBN: 92 5 104835 5. ftp://ftp.fao.org/docrep/fao/005/y4252e/y4252e.pdf

[2] **J. Roca, I. Parilla, A. Bolarin, E.A. Martinez and Heriberto Rodriguez-Martinez,** Will AI in pigs become more efficient?, 2016, Theriogenology, (86), 1, 187-193. http://dx.doi.org/10.1016/j.theriogenology.2015.11.026

[3] **J. Roca, WLWJ. Broekhuijse, I. Parilla, H. Rodriguez-Martinez, EA. Martinez and A. Bolarin,** Boar Differences in Artificial Insemination Outcomes: Can they Be Minimized?, Reprodion Domestic Animals 2015;50:48-55.

[4] **D. Waberski, A. Riesenbeck, M. Schulze, K. Fritz Weitze, L. Johnson,** Application of preserved boar semen for artificial insemination: Past, present and future challenges, 2019, Theriogenology, (137), 2-7.

[5] **L. Spjuth, A. Johannisson, N. Lundeheim, H. Rodriguez-Martinez**, Early pre-pubertal exposure to low dose oral di(2-ethylhexyl) phthalate does not affect sperm plasma membrane stability, acrosomal integrity or chromatin structure in the post-pubertal boar. Theriogenology 2007; 68: 186-95.

[6] **M. Schultze, K. Rudiger, D. Waberski,** Rotation of boar semen doses during storage affects sperm quality. Reproduction Domestic Animals 2015;50:684-7. https://doi.org/10.1016/rda.12532

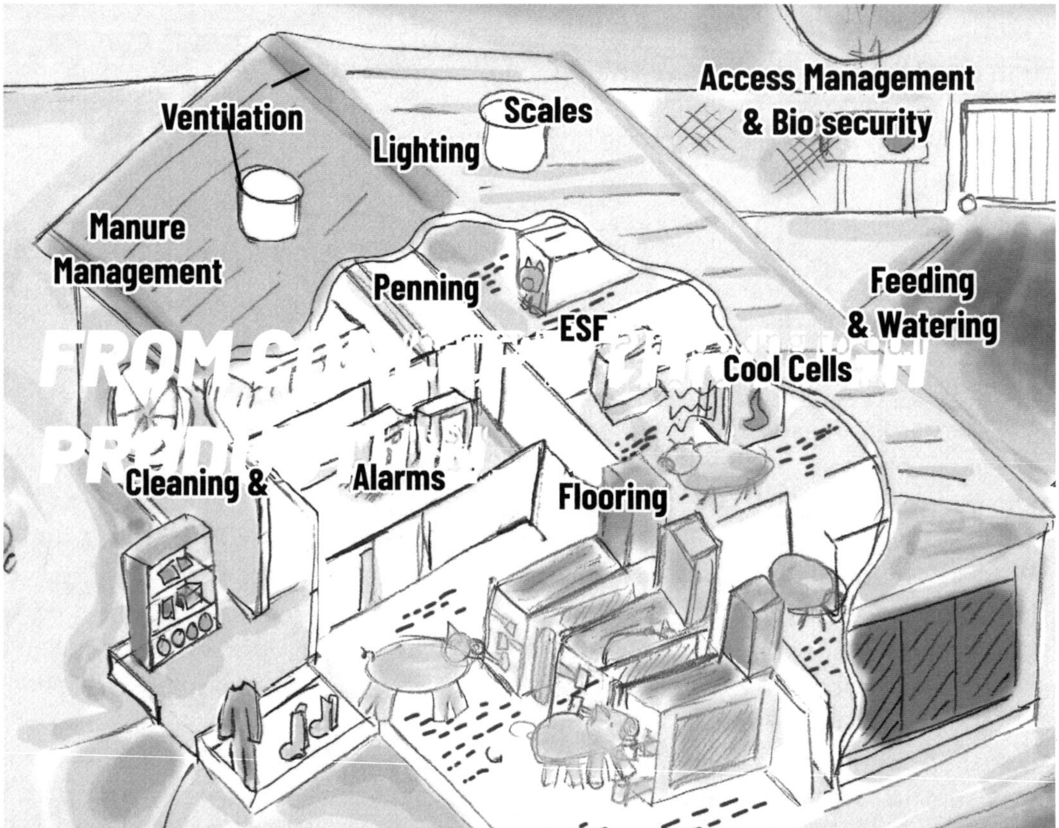
Ventilation
Scales
Lighting
Access Management
& Bio security
Manure
Management
Penning
ESF
Feeding
& Watering
Cool Cells
Cleaning &
Alarms
Flooring

Digital Farm & Technology Insight

David Speller

OPTIfarm
Chesterfield, UK

1. Introduction

Undoubtedly terms like digital farming, precision livestock farming, technology, and data insights, have all become at the top of many livestock conference agendas. The terms offer so much yet often deliver so little when we try to carry them on to the livestock farm. Indeed, it appears arable and crop farming is accelerating away from us with self-steering, precision applications, GPS mapping and soon to come self-driving or driverless tractors, fruit pickers, etc. This manuscript will lay out the foundations for the breakout session to be delivered exploring 'Digital Farm and Technology Insights' with a focus on livestock enterprises. IN the breakout session we will explore specific examples and understand case studies but as a prelude this manuscript lays the foundation to the thinking.

2. What are the Trends

The past 10 years plus the livestock sectors have been focusing on gathering farm and production data such as environmental sensor data, feed and water consumption data and performance data. The aim was to capture it, store it and conduct some analysis of it. This will surely lead to an increased understanding of our businesses, better decision making on farm and increased livestock productivity, health and welfare. This trend has definitely set businesses on a journey but not so many businesses have seen the real generation of tangible extra profits related solely to the digital and technology projects they have implemented. Whilst the data was being gathered the technology companies, fueled by inspirational scientists, have developed more and more exciting technologies offering even greater insights into the world of farmed animals not previously possible to explore. Most recently we have seen a focus on using technology to conduct the same evaluations a skilled stock person may conduct, focusing on sight, sound and smell, with a lot of activity around visual data. Some technologies have been specifically developed for the agricultural implementation such as camera weighing systems or microphones listening to animals whilst others have been taken from other sectors and adapted for agriculture, such as an e-nose originally designed for human health or robotics from the nuclear industry.

We now appear at a point where the possibilities of technology know no bounds, as challenges are overcome around internet connectivity and the ability to export data from farms, we see smarter solutions to process and compress data on farm before exporting via cellular networks or satellites, including the new low orbit, Starlink satellite solution.

We have now reached a place where we appear full of data, with more different data sets arriving daily and an ever-growing desire to find ways to use this data, from driving productivity, validating supply chains and more recently answering the questions around sustainability. However, there are still few ways of turning any of this data into real life business actions and without this the ability for the digital and technology industries to deliver value to livestock farms is being challenged.

3. What are the Problems Being Solved

Although the specific problems in different livestock sectors vary there are some common themes. Primarily we see technology being developed to bring new insights and understanding, reduce the demand on labour or support the labour present and compile data and data analytics to enhance the prediction of animal health challenges or improving planning of product harvesting. Examples of this include monitoring animal activities to understand behaviour and activity levels, robotics to replace or support human tasks such as milking, poultry or swine weight recording to better predict processing weights and sound analysis predicting respiratory diseases in swine.

4. Who is Driving the Development and Uptake of Digital Solutions

Whilst I have referred to technology companies, fueled by interesting science, delivering digital and technology solutions it is also worth noting that most of the digital and technology solutions we see globally are originating from a start-up culture, either as a spin out of an established business/ academic institution or as a standalone start up. By their very definition start-ups require early investment funds or early adopters willing to fund the realization of a concept. This process alone selects those ideas that suit investments and early adopters. We have definitely seen a real desire for such start-up facilitators to be attracted to terminology such as artificial intelligence, machine learning, SaaS, reoccurring revenue, etc. This has really driven most developments to answer the call of their investors and have to offer new innovative business models, new solutions and a drive to finding a solution that are believed to meet the mutual needs of a global livestock sector.

In the last few years we have now seen the larger supply companies in the livestock sectors acquiring innovation rather than incubating it internally and they are also showing a desire for the new digital and technology solutions to support their legacy products. It appears that we may be on the cusp of seeing digital and technology solutions and their insights rolled out on scale fueled by an existing supply chain that have a need to find new ways to sell their products. Some solutions such as robotic milking machines, automated environmental controls, production metrics forecasting, such as weight, have seen a good roll out but really there are few digital and technology solutions that have seen rapid global scaling to date. The sector is still wondering who is the customer for their solutions, is it the farmer, the processor or the allied supply industries?

5. What Business Environment Do We Need to be Able to Adopt These Solutions

Digital insights and technology do have the potential to support business improvements and productivity gains but there is an issue in relating the value of the gains with the purchaser of the technology. An example of this might be the use of insights and technology to better deliver uniformity in a batch of swine delivered to the processor but if the commercial contract does not reward the farmer for the improved uniformity the processor has an opportunity to gain processor yield improvements and processing speed improvements without any costs incurred whilst the farmer has the cost to bear for the digital solution. Where we might see the reverse of this is where premiums maybe available for "No Antibiotics Ever" reared swine and insights and technology has the ability to support this production technique thus making the case for investment in digital and technology a more simple decision.

6. Are We Embracing Digital Solutions and Technology Fast Enough

There are a variety of rates of uptake for digital solutions and technology across different livestock sectors and different regions of the world. Ideas and offerings have been around in a 'proof of concept' format for many years and so there may be an argument that uptake is too slow However, availability of commercially ready offerings is more recent and it appears many are still working out the return on investment for these solutions and which are the right solutions for solving the right challenges. What we need to see now is the growing acceptance and use of these digital solutions as not only does this endorse the digital and technology solutions themselves to the industry but also the learnings of the companies offering these solutions will grow much faster as they see and learn from their solutions being deployed in the field on scale.

7. What are Tomorrows Needs Likely to be From the New Technologies

Every company is now committed, focused and addressing the demands of their own sustainability commitments. Whilst much of sustainability compliments good business practices with resource efficiency and output optimization what will also be needed moving forward is a way to validate any statements made and this is where digital and technology insights can really support a business on all elements of sustainability, social, economic, and environmental. Capturing sustainability related data and validating company commitments will become a growing area that we will see more of.

In addition, labour resources in all parts of the livestock sector are under pressure everywhere around the globe and so we will need to find solutions to meet the challenges around labour availability. This may be more technologies such as robotics removing the tasks humans are expected to do, new data streams better informing and supporting less experienced labour resources or technology solutions that allow new work patterns on farms, such as shift working, shared responsibilities, etc.

8. Conclusions

By exploring the topic in a broader context than just individual case studies and their success or failure it is possible to understand the development of digital farm and technology offerings in many livestock sectors globally. Their uptake can be seen as an evolution not a revolution. However, there has also been a need for other critical elements such as remote, rural, internet connectivity and the declining cost of hardware to catch up with the aspirations of the livestock sector. In many farm enterprises the margins between profitability or not are narrow and so solutions must be cost effective, offer strong and reliable return on investments and meet the present days needs. All this from the coming together of academia, start-ups and large corporates means this is a hugely interesting and exciting time for all involved.

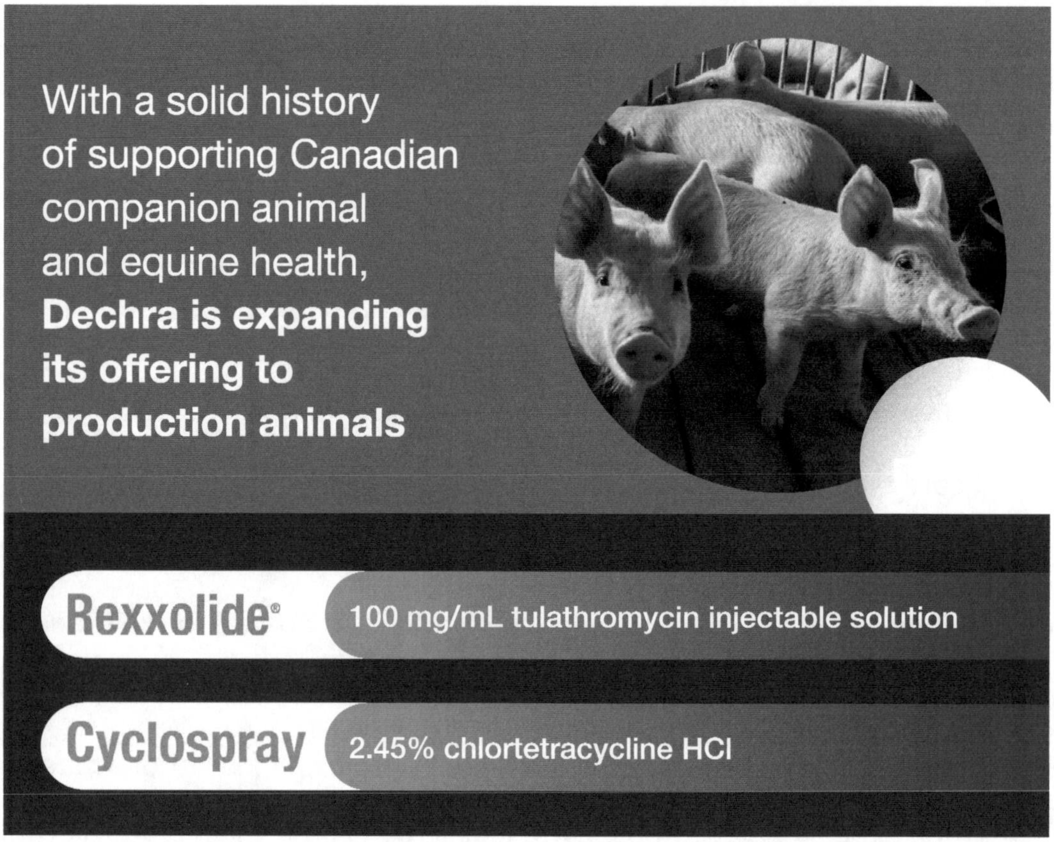
With a solid history
of supporting Canadian
companion animal
and equine health,
Dechra is expanding
its offering to
production animals
Rexxolide®
100 mg/mL tulathromycin injectable solution
Cyclospray
2.45% chlortetracycline HCl

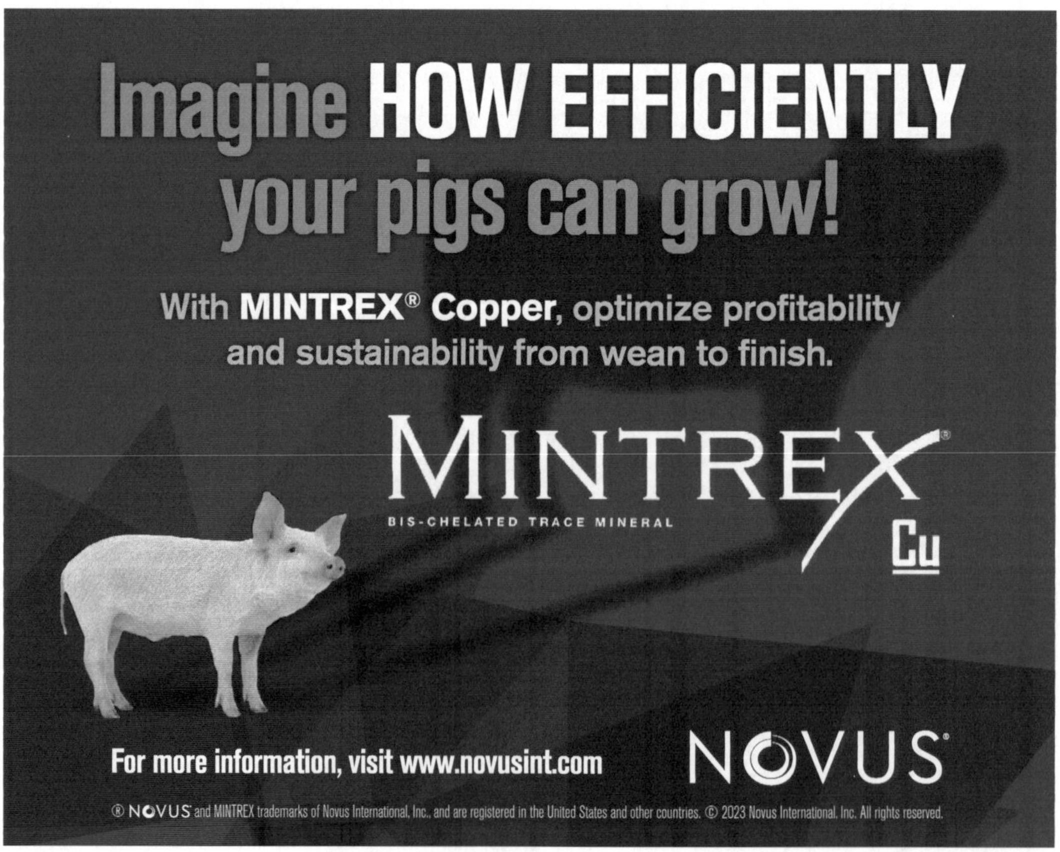
Imagine HOW EFFICIENTLY
your pigs can grow!
With MINTREX® Copper, optimize profitability
and sustainability from wean to finish.
MINTREX®
BIS-CHELATED TRACE MINERAL
Cu
For more information, visit www.novusint.com
NOVUS®
® NOVUS and MINTREX trademarks of Novus International, Inc., and are registered in the United States and other countries. © 2023 Novus International, Inc. All rights reserved.

Improves piglet and sow health
Jefo's specific blend of protected organic acid for piglet and sow, will help improve gut health and reproductive performance.
Increases growth rate and lowers incidence of diarrhea at weaning
Improves litter size and body condition
Reduces vulval discharge and UTI
Jefo
For more information, contact your specialist :
Rhod Ross
North Alberta / Saskatchewan
(403) 702-6709 | rross@jefo.ca
Ron Jones
Alberta / Saskatchewan
(403) 804-1205 | rjones@jefo.ca
Alli Garant
Saskatchewan
(306) 230-3800 | agarand@jefo.ca
Alex Lima
Ontario / Manitoba
(416) 557-5751 | alima@jefo.ca
Dominic Frappier
Quebec
(450) 278-0661 | dfrappier@jefo.ca
Chris Gwyn
North America
(519) 301-8234 | cgwyn@jefo.ca

Evaluation and Application of Technology in Swine Production Systems

David S. Rosero

The Hanor Company
4005 E Owen K Garriott Rd, Enid, OK 73701
Email: drosero@hanorusa.com

1. Introduction

Swine production systems are rapidly adopting smart farming technologies to address the increasing pressure of higher production costs and the limited access to qualified labor. In addition, our society is increasingly demanding an enhanced capacity for food production while improving efficiency in the use of resources in efforts to mitigate climate change (FAO, 2013).

We anticipate that smart farming tools will continue to be implemented in large commercial systems as they continue to be proven accurate, reliable, and cost-effective. We expect that the adoption of innovative technologies into swine barns will drive substantial improvements in animal well-being, animal growth and performance, and maximization of revenue capture. This paper describes emerging innovations and key aspects when evaluating technology accuracy, reliability, and implementation models in large commercial systems.

2. Emerging Smart Farming Technologies

The rapid development and constant improvement of Internet-of-Things devices, edge computing, and artificial intelligence models allows for a continuous monitoring of animal health status and response to medical treatment; improvement of animal well-being, especially when the animal is most vulnerable; and monitoring of growth and performance to maximize revenue capture (Vranken and Berckmans, 2017). Ultimately, continuous real-time data captured by smart devices must be integrated, processed (statistical analysis, data visualization, artificial intelligence) and presented to end-users via friendly interfaces for decision making.

Significant innovations are emerging and proposing to solve simple and complex tasks in swine barns. Special attention has been given to computer vision technologies because of the low cost of cameras and the advancements in edge computing systems that automatically extract, analyze, and understand complex problems from a single image or a sequence of images. Huang (1996) defined computer Vision as a scientific field that seeks to automate tasks beyond the capacity of the human visual system. Applications of computer vision technologies in agriculture include the monitoring of growth, disease identification and prevention, automatic harvesting, compliance with animal welfare (Arulmozhi et al., 2021). In swine production systems, computer vision technologies have been applied to monitor animal behavior, counting of pigs, and estimation of body weight (Tokach et al., 2021; Franchi et al., 2022). Real-time access to this information is critical to assess animal survivability and welfare, and to accurately predict market weights and profitability.

3. Application of Computer Vision in Swine Barns

Recent advancements in computer vision methods and the use of robust equipment have overcome field limitations such as diverse lighting conditions, different backgrounds (color and shapes of walls and floors), and multiple object barriers, including dust and insects in contact with lenses of cameras. Pezzuolo et al. (2018) reported that depth model cameras accurately (99%) predicted the weight of pigs from 13 to 104 lb. Moreover, Abner et al. (2021) validated the accuracy of smart cameras (PigVision® Asimetrix Inc; Durham, NC) and compared them to human estimations, a walk-across platform scale (CIMA; Correggio, Italy) and a calibrated livestock scale. Accuracy of individually weighed pigs was 98.2% for the walk-across scale (did not register a weight for six pigs); 88.2% for human observation, and 96.6% for PigVision®.

Recently, various prototypes of pig counting technologies were developed by researchers from Kansas State University (KSU) in collaboration with engineers and technology experts from Amazon Web Services (AWS), and production specialist from JBS (Tokach et al., 2020). The first prototype resulted in a wide range of pigs counted, with the highest accuracy of 94.6%. Accuracy results of the second prototype ranged from 76 to 99.1% for small group of pigs (50 to 150 pigs). Prototype 3 was developed for field implementation leveraging learning and results from the previous prototypes. Overall, 17,000 pigs were counted. The accuracy average was 79%, ranging from 50 to 100%.

4. Field Validation of Computer Vision Technologies in Swine Barns

We propose that field evaluation of smart farming technologies must include assessments of accuracy and reliability. Moreover, validation tests should be conducted under field conditions. Our evaluation process of the technology PigVision® that estimates weight of pigs was conducted in a research-commercial finish barn (100 pens, capacity for 2,400 pigs) in Iowa by using a total of 32 cameras. Estimation of weights by cameras were compared to weights from a built-in pen scale every 2 months for a period of more than 2 years. Average accuracy has improved overtime from < 90% during the initial tests to > 96% in the most recent cycles.

Field validation of cameras overtime allowed the assessment of reliability of the equipment by measuring the uptime of cameras. We defined uptime of cameras as the percentage of daily weight estimations reported to end users relative to the number of cameras placed in the barn. During the first cycle of production (Apr to Jul 2020), the uptime averaged 97.2%. The uptime decreased to 85.6% in the next cycle (Aug to Dec 2020). Changes in cameras, internet reliability and periodicity in lenses cleaning resulted in improving the uptime to > 98.5% during the last 2 cycles tested (Feb to Oct 2022).

5. Conclusions

Swine production systems and their biological processes are extremely complex to attempt replacing human husbandry by technology, but rather we expect that emerging smart farming systems will augment the farmer's capacity. The complexity of problems in farms require a major shift from our conventional methods of production. Technology smartness for continuous monitoring of animal growth will drive substantial improvements to support animal welfare and to monitor changes in health status and production efficiency.

References

Abner, V. A. (2021) Accurate and rapid weight assessment of finishing pigs. Thesis. Available at: https://repository.lib.ncsu.edu/bitstream/handle/1840.20/39036/etd.pdf Accessed 20 November 2022.

Arulmozhi, E., A. Bhujel, B. E. Moon, H. T. Kim (2021) The application of cameras in precision pig farming: an overview for swine-keeping professionals. Animals 11:2343.

FAO (2013) Climate-smart agriculture sourcebook. Available at: http://www.fao.org/climate-smart-agriculture-sourcebook/en. Accessed 29 November 2020.

Franchi, G. A., J. D. Bus, I. J. Boumans, E. A. Bokkers. M. B. Jensen, L. J. Pedersen (2022) Estimating body weight in conventional growing pigs using a depth camera. Smart Agriculture Technology 3:2023.

Huang, T. (1996) Computer Vision: Evolution and promise. 19th CERN School of /Computing. Geneva: CERN. pp. 21–25.

Pezzuolo, A., M. Guarino, L. Sartori, L. A. González, and F. Marinello (2018) On-barn pig weight estimation based on body measurements by a Kinect v1 depth camera. Computer and Electricity in Agriculture 148: 29-36.

Tokach, M. D., J. C. Woodworth, A. B. Lerner, K. F. Coble, S. Blazevich, J. M. DeRouchey, S. S. Dritz, R. D. Goodband, K. Duggal (2020) Developing an automated, real-time pig counting technology. NPB Research Report # 19-080.

Vranken, E., and D. Berckmans (2017) Precision livestock farming for pigs. Animal Frontiers 7:32-37.

Colostrum Management and Piglet Survival

Kara R. Stewart

Purdue University, West Lafayette, IN 47907
Email: krstewart@purdue.edu

Advancements in genetic selection has led to highly prolific sows producing piglets with economically important carcass traits. In the US and Canada, total born has increased by 3 piglets or more since 2004 (PigChamp, 2021). As litter size has increased, variation in birth weight has also increased (Milligan et al., 2002) as well as the emergence of the low birthweight phenotype sow (Foxcroft et al., 2009). Piglets born with a light birthweight have decreased survival, which has driven pre-weaning mortality (PWM) up in the US and Canada from approximately 12% in 2004 to 15% on average in 2021, with some farms reaching over 20% mortality (PigChamp, 2021). The vast majority of these piglet deaths are multifactorial and occur during the first three days after birth (Panzardi et al., 2013, Muns et al., 2016. Lay et al., 2002). Approximately 11% of all piglets born weigh <1.0kg and have a PWM of approximately 36-40% (K. Stewart, unpublished), which accounts for ~29% of total mortality. While the likelihood of survival increases as piglets birth weights increase above 1kg, there are many more piglets born in the medium and heavy weight categories. So, piglets in the medium birthweight category (>1.0-1.6kg), may have a lower PWM of about 10-16%, but they account for 55% of all the piglets born. Therefore, their overall contribution to total mortality is 50% (Table 1).

Table 1. Commercial data showing contribution of birthweight to overall mortality. Data from The Maschhoff's farm with an overall preweaning mortality of 13.5 %.

Birth Weight (kg)	NBA	% of Total born	PWM, %	Number Dead Piglets	Percent of Total Mortality	Birthweight Category
<1.0	417	10.9%	36.2%	151	29%	Light – 10.9% of all piglets born 29% of total mortality
1.0-1.2	517	13.5%	16.1%	83	16%	Medium – 55% of all piglets born 50% of total mortality
1.2-1.4	748	19.5%	12.2%	91	17.6%	
1.4-1.6	832	21.7%	10.1%	84	16.2%	
1.6-1.8	721	18.8%	8.7%	63	12.2%	Heavy – 15.7% of all piglets born 21% of total mortality
1.8-2.0	409	10.7%	6.8%	28	5.4%	
2.0-2.2	154	4%	7.8%	12	2.3%	
>2.2	41	1.1%	14.6%	6	1.2%	

One of the driving factors leading to decreased survival in piglets is colostrum intake (Le Dividich et al., 2005). As litter size has increased, sow's colostrum production has not, resulting in each piglet consuming approximate 42g less of colostrum for every additional piglet added to the litter (Devillers et al., 2011). Piglets that consume more than 200g of colostrum in the first 24h after birth have increased chances of survival to weaning (Devillers et al., 2011, K. Stewart, unpublished Table 2). As you can see from Table 2, piglets that consume <200g of colostrum are typically the light birthweight piglets and have a high PWM of 39%. The medium weight piglets (>1.0-1.5kg) are consuming 250-350g of colostrum in the first 24h after birth and have much lower preweaning mortalities. While the 200g amount has been determined for increased survival in the preweaning period, the amount of colostrum needed to survive in nursery and finishing or to become a productive replacement sow may be different. For example, gilts that gain

≥0.22g/d during the preweaning period have >45% of remaining in the herd through parity 1 (Knauer, 2016). To optimize age at puberty in gilts, they must have a pre-weaning average daily weight gain (ADG) of 0.26g/d (Vallet et al., 2016). If we look back at Table 2, gilts that achieve this ADG for these two scenarios consume 250-400g of colostrum in the first day of life. Therefore, while 200g of colostrum may be required to survive, it is likely that 300g or more are needed to thrive.

Table 2. Preweaning mortality rates and growth of piglets as colostrum intake increases. Data unpublished (K. Stewart) from 648 piglets from Purdue University farm.

Colostrum Intake, g	% of piglets	Average BW, kg	ADG birth to wean, kg/d	PWM, %
150-200	2.8	0.70	0.187	39
200-250	11.9	0.92	0.204	7.8
250-300	21.8	1.16	0.222	6.4
300-350	30.0	1.4	0.245	5.6
350-400	22.3	1.63	0.264	6.9
400-450	8.6	1.9	0.269	7.1
>450	2	2.1	0.296	7.7

It is likely that roughly 60-70% of piglets born are consuming enough colostrum to survive and thrive. Risk factors for piglets to have low colostrum intake are piglets born late in the birth order, piglets born to large litter sizes, and piglets unable to regulate their body temperature after birth, resulting in missed nursing events. Management practices such as crossfostering, split suckling, drying piglets, and supplementing piglets have been used to increase colostrum consumption by piglets, with most of these efforts focusing on the light birthweight piglets. While many of these management practices are used today, few have shown direct benefits to increased colostrum consumption when evaluated in controlled trials. It should be noted, that the vast majority of the controlled trials are on very few numbers of animals with mixed results shown to increase preweaning growth and survival. Anecdotally, adoption of some of these management techniques can reduce PWM. Therefore, we can assume that there isn't a single management practice that will work for every farm under every management strategy.

What can be seen however, is that increased labor in the farrowing barn whether they are focused on drying piglets at birth, supplementing piglets, or conducting split suckling, will save more piglets. A controlled trial investigated the impacts of supervision during farrowing and for 3d after and found an additional weaned pig per litter in supervised litters compared to unsupervised (Holyoake et al., 1995) due to a decrease in stillborns and live born mortality. Directing labor to the farrowing house is a beneficial endeavor to save piglets.

Labor comes as a high cost, so there is great potential for the future of technology to help identify piglets that have not consumed enough colostrum. Piglets that consume low amounts of colostrum struggle to regulate their body temperature at 24h of age and have a slower growth rate, all of which has potential for technology to identify. Additionally, genetic selection for increased teat numbers has been shown to increase colostrum output by 324g for each additional teat (Weigert and Knauer, unpublished), enough for an additional piglet to survive and thrive.

The dairy industry has developed a scoring system to identify calves that have received adequate colostrum based on transfer of immunoglobulins to the calf. A survey of the success of passive transfer of immunoglobulins among dairies found that about 80% of calves receive adequate colostrum, and this is increased on smaller farms and when colostrum is provided within 4 hours of the birth of the calf (Beam et al., 2009). A goal of the swine industry must be to figure out methods how to identify piglets that have not received enough colostrum and methods to increase colostrum ingestion by these piglets. Colostrum's role is multifactorial and contributes to thermoregulation, immune function, nutrition, and final maturation of body organs. Therefore, its role in the piglet goes beyond the farrowing house and must be optimized in the first day of life.

References

Beam, A. L., et al. (2009). "Prevalence of failure of passive transfer of immunity in newborn heifer calves and associated management practices on US dairy operations." J Dairy Sci 92(8): 3973-3980.

D.C. Lay, J., R.L. Matteri, J.A. Carroll, T.J. Fangman, T.J. Safranski (2002). "Preweaning survival in swine." Journal of Animal Science 80(E. Suppl. 1): E74-E86.

Devillers, N., et al. (2011). "Influence of colostrum intake on piglet survival and immunity." Animal 5(10): 1605-1612.

Foxcroft, G. R., et al. (2009). "Prenatal programming of postnatal development in the pig." Soc Reprod Fertil Suppl 66: 213-231.

le Dividich, J., J.A. Rooke, P. Herpin (2005). "Nutritional and immunological importance of colostrum for the new-born pig." Journal of Agricultural Science 143: 169-485.

Magnabosco, D., Bernardi, M.L., Wentz, I., Cunha, E.C.P., Bortolozzo, F.P. (2016). "Low birth weight affects lifetime productive performance and longevity of female swine." Livestock Science 184: 119-125.

Milligan, B. N., et al. (2002). "Neonatal-piglet weight variation and its relation to pre-weaning mortality and weight gain on commercial farms." Prev Vet Med 56(2): 119-127.

Muns, R., et al. (2013). "Piglet behavior as a measure of vitality and its influence on piglet survival and growth during lactation." J Anim Sci 91(4): 1838-1843.

Panzardi, A., et al. (2013). "Newborn piglet traits associated with survival and growth performance until weaning." Prev Vet Med 110(2): 206-213.

Vallet, J. L., et al. (2016). "Litter-of-origin trait effects on gilt development." J Anim Sci 94(1): 96-105.

Setting the standard
Hypor
Hypor Magnus
HIGH QUALITY PIGS
EFFICIENT AND UNIFORM FINISHERS
MORE QUALITY MEAT
hypor.com
HENDRIX GENETICS

Assessing Economic Opportunity of Improving Mortality Rate in Breed-to-Wean Swine Production

Lee Schulz

Iowa State University

One of the challenges facing many sow farms is high death loss. The Pig Computerized Health and Management Program, or PigCHAMP, is a database that includes information from nearly 300 farms. PigCHAMP's website provides publicly accessible benchmark summaries which have shown an increase in sow mortality rate from 8.12% in 2012 to 14.86% in 2021.[1] In 2021, the upper 10 percentile of herds for sow mortality had an average death rate of 21.30%. On the other hand, the lower 10 percentile for sow mortality had a death rate of 7.30%. These values clearly illustrate the extremes that can be seen on individual farms for sow mortality and the potential to improve.

Improving sow mortality, and pre-wean mortality which is a secondary focus of this analysis, results in greater efficiency and producing more pigs, which can increase potential profits. In most, if not all cases, there will be some cost associated with reducing mortality rate, so producers need to weigh the costs against the potential economic benefits. Individual farms may have different methods to reduce mortality and the costs will also vary. For example, costs could include additional labor, health and management interventions, or facility improvements depending on the causes of mortality. This analysis focuses on the benefit of improving mortality on an individual operation. By knowing the benefit, one can back into indifference points to guide decision-making and help identify an optimal level of mortality for an operation at a particular point in time.

Quantifying the pig production effect and economic returns associated with improving sow mortality is complicated because of the numerous interacting factors. For example, if sow mortality decreases, do total sow numbers decrease or stay the same and are there other production implications on a particular operation? Because of the numerous factors to consider, any economic analysis will require several assumptions. The validity of these assumptions must be tested within an operation and how the model projects actual performance. A sensitivity analysis with some of these key factors is an important part of the economic analysis to help determine what factors are most important.

Method of Economic Analysis

Projected annual budgets of an operation that is in steady-state or on-going, not one that is starting up, are used to compare costs and returns for alternative sow mortality rates. Costs in the budget reflect full economic costs rather than cash flow costs. The annual budget model uses annual averages for production efficiencies and mortalities and does not account for variation that may occur during seasons (i.e., summer versus winter) or from one farrowing group to the next but does provide a basis for assessing the economic opportunity for improving mortality rate over time.

The model uses a current and improved mortality rate and other related inputs to quantify the economic effects of changes in mortality. Key inputs that affect net income changes are highlighted in this Information File. An accompanying Decision Tool spreadsheet, allows users to employ their own production and budget inputs to assess their individual situation. A sensitivity analysis that compares net income changes due to changes in sow mortality and pre-wean pig mortality and changing costs or prices is also an important component of the tool.

Assumptions in the Economic Analysis

Numerous assumptions were made to construct the model. The following are some of the key assumptions that impact the analysis.

- The model requires inputs for current mortality rate, litters per mated female per year and average sow inventory. These should be as accurate as possible and based on record information over a period of steady production.
- It is assumed that most breed-to-wean operations use breeding targets that maximize litters for the space available. However, the potential litters from sows that die between breeding and farrowing cannot be replaced. To model this loss, litters per mated female per year change by 0.637%

per 1% of sow mortality change that occur after breeding and before farrowing. The 0.637% factor is based on the average change in litters per mated female per year of production simulations with different breeding and farrowing schedules, farrowing rates, and mortality rates.

- Average sow inventory is assumed not to change with mortality rate. With lower mortality rates, fewer sows may be needed to produce the same number of pigs, but it is assumed sow numbers will be based on facility size and operating at full capacity.
- It is assumed facilities and space are not limiting and are available for all gilts, sows, and farrowings.
- For simplicity, the parity of individual sow mortalities is not considered in this analysis. It is recognized that several studies have analyzed the economic benefit of sow longevity and the productivity advantages of mid-parity sows versus gilts or older sows, but it is not modeled here. Along these same lines, wean-to-finish performance of progeny by parity is not considered.
- Replacement gilt cost changes 1% for each 1% in change in total sow mortality. It is assumed sow mortalities are replaced by gilts that are ready to be bred. In other words, a supply of developed gilts is available to replace all sow mortalities. The cost input for replacement gilts should reflect the cost of developing the gilt and the additional cost of gilts in the gilt pool that do not enter the breeding herd. The inputs of culling rate and mortality rate added together determine the replacement rate in the model.

- Cull sow income changes by 1% for each 1% change in total sow mortality. It is assumed cull sow sales will increase with lower mortalities. All cull sow values are considered equal. The user can input a culling rate which is independent of mortality rate.
- The amount of feed consumed is assumed not to change as mortality rate changes. An annual feed quantity per sow is used in the budget and sow numbers are considered static.
- The selling price of a weaned pig is constant across all scenarios. In other words, it is assumed that sow mortality has no impact on progeny value.

In addition to the assumptions listed above, there are key inputs that affect incomes and costs across alternative sow mortality rates. The income changes related to sow mortalities are weaned pig income and cull sow sales.

The main inputs for weaned pig income are sow mortality rate after farrowing and before breeding, current litters per mated female per year, pigs born alive per litter, pre-wean mortality rate, and weaned pig price.

The main inputs for cull sow income are total sow mortality rate, cull sow weight, and cull sow price.

The primary cost change that occurs with changes in sow mortality rate is increased replacement gilt costs. The key inputs are the cost per head of replacement gilts, total sow mortality, and culling rate.

The number of pigs weaned per sow per year is a commonly cited measure of breeding herd productivity, and pre-wean mortality is one of its principal determinants. Pre-wean pig mortality reduces the number of pigs weaned and results in lower income as pre-wean mortality rates increase. If pigs born alive per sow increases due to changes in sow mortality rate, the pre-wean mortality rate would need to decrease to compare the same number of pigs lost due to pre-wean mortality.

Other variable and fixed costs in the budget do not change for the operation as sow or pre-wean mortality rate changes. Those cost may be different on a per pig weaned basis in comparison as total costs are divided by different numbers of pigs.

Example Analysis

The Ag Decision Maker Decision Tool spreadsheet, has been developed to aid in analyzing the economic opportunity for improving mortality rate. Example inputs and outputs from the Decision Tool will be used here to show the economic opportunity of reducing sow mortality by one percentage point.

Inputs

Inputs used for this analysis were estimated from industry averages, where possible, except for the number of sows. A herd size of 2,500 sows is used. Total sow mortality of 14% is compared to 13%. The mortality percent after farrowing but before breeding is assumed to be 50% of total mortalities based on industry reports.[2] So in this example, 7% as compared to 6.5%. Pre-wean mortality is 15% for current and changed to 14.95% in the improved scenario to maintain the same number of pigs lost to pre-wean mortality.

The current litters per mated female per year is 2.20 which increases to 2.207 due to a one-half percentage point improvement in sow mortality during gestation. Pigs born alive per litter and pigs weaned per sow per year is 13.50 and 25.25, respectively, and pigs weaned per sow per year increases to 25.34 with the improvement in sow mortality. The culling rate is 45% and cull sows are assumed to weigh 450 pounds. Table 1 shows the remaining information used for the example analysis.

Table 1. Production price and cost inputs.

Table 1. Production, price, and cost inputs		
Price inputs		
Weaned pig price	$58.00	per head
Cull sow price	$48.00	per cwt.
Cost of developed replacement gilt ready to breed	$375.00	per head
Variable inputs and costs		
Feed cost	$ 0.15	per lb.
Feed processing and delivery cost	$18.00	per ton
Pounds of feed	2,350	per sow per year
Labor cost	$15.00	per hour
Labor hours	7.00	per sow per year
Veterinary and health cost	$40.00	per sow per year
Semen cost, genetic fee	$30.00	per sow per year
Marketing and professional fees	$4.00	per sow per year
Utilities and fuel cost	$25.00	per sow per year
Machinery, facility and equipment repairs	$22.00	per sow per year
Other variable costs	$10.00	per sow per year
Fixed costs		
Machinery, facilities, and general overhead	$170.00	per sow per year
Taxes and insurance	$10.00	per sow per year
Legal and accounting	$11.00	per sow per year
Other fixed costs	$0.00	per sow per year

Outputs

A partial budget is divided into two sections designed to separately present important components of the analysis. Table 2 shows the current and improved sow mortality rates compared to zero mortality on a per sow per year and per operation per year basis. While it may seem unrealistic to consider operations that have zero sow mortality, these scenarios are included in the analysis for comparison purposes and to highlight the cost of mortality. It is important to keep in mind that there is an optimal level of mortality, which is not zero because the marginal benefit of reducing, or attempting to reduce, mortality below a certain point is less than the marginal cost. In other words, determining the optimal death loss is about determining the percentage of mortality at which the cost of saving additional sows is no longer offset by the added revenue those sows will bring in. Pre-wean mortality rates for the improved scenario are adjusted so the number of pigs that die before weaning is the same and there is no difference in the current and improved scenarios due to pre-wean mortality. Table 2 also summarizes the impact of the one percentage-point change in total sow mortality from 14% to 13% and a one-half percentage point change in sow mortality from farrowing to breeding. The results are presented on an annual per sow and per operation basis. The improvement in sow mortality results in an $11.40 per sow per year increase in net income. About 50% of the net income change is due to increased pigs weaned, 20% is additional cull sow income, and 30% is reduction in gilt replacement cost.

Table 2. Partial budget for marginal changes.

Table 2. Partial budget for marginal changes						
	Compared to 0% mortality				**Improved mortality rate compared to current rate**	
Sow mortality economic impact	**Per sow per year**		**Per operation per year**		**Effect of 1% change in total sow mortality and 0.5% change in sow mortalities during gestation**	
	Current	**Improved**	**Current**	**Improved**		
Revenue change	**14% and 7%**	**13% and 6.5%**	**14% and 7%**	**13% and 6.5%**	**Per sow per year**	**Per operation per year**
Pig value (without pre-wean mortality impact)	$(76.81)	$(71.32)	$(192,027)	$(178,311)	$5.49	$13,716.20
Cull sow income	$(30.24)	$(28.08)	$(75,600)	$(70,200)	$2.16	$5,400.00
Cost change						
Replacement gilt cost	$52.50	$48.75	$131,250	$121,875	$(3.75)	$(9,375.00)
Net income change due to sow mortality	**$(159.55)**	**$(148.15)**	**$(398,877)**	**$(370,386)**	**$11.40**	**$28,491.20**
Pre-wean mortality impact	Compared to 0% pre-wean mortality				Effect of 0.048% change in pre-wean pig mortality	
Pig value	$(258.39)	$(258.39)	$(645,975)	$(645,976)	$(0.00)	$(1.01)
Net of sow and pre-wean mortality	**$(417.94)**	**$(406.54)**	**$(1,044,852)**	**$(1,016,362)**	**$11.40**	**$28,490.20**

The total cost per sow lost can also be calculated. In the example, a one percentage point change in sow mortality is equal to 25 sows. The operation income change of $28,490 divided by 25 sows is equal to $1,140 per sow lost in this example. For the purposes of this calculation, recall, the mortality percent after farrowing but before breeding is assumed to be 50%. But, if more sows die after farrowing a litter and before breeding the cost is less than if more sows die during gestation.

While a partial budget shows the marginal impact on the operation, an enterprise budget shows the impact to the bottom line. Income over variable costs and income over total costs, for example, may be more useful for a manager than net income change. Using the assumptions and production, price, and cost inputs in this example, Table 3 summarizes the full enterprise budget on an annual per sow, per pig, and per operation basis.

Table 3. Annual Enterprise budget for bottom line changes, summary.

Table 3. Annual enterprise budget for bottom line changes, summary

	Per sow		Per pig produced		Per operation	
	Current	Improved	Current	Improved	Current	Improved
Gross Income	$1,574.41	$1,582.06	$62.37	$62.43	$3,936,025	$3,955,140
Total Variable Costs	827.75	824.00	32.79	32.52	2,069,375	2,060,000
Total Fixed Costs	191.00	191.00	7.57	7.54	477,500	477,500
Total Costs	1,018.75	1,015.00	40.35	40.06	2,546,875	2,537,500
Income Over Variable Costs	746.66	758.06	29.58	29.92	1,866,650	1,895,140
Income Over Total Costs	555.66	567.06	22.01	22.38	1,389,150	1,417,640

For this example operation, a one percentage point improvement in total sow mortality and a one-half percentage point decrease in sow mortality after farrowing and before weaning would result in a $28,490 increase in annual income over total costs. Different assumptions or inputs would change the outputs.

For this operation, sow mortality reducing strategies that would cost less than $28,490 would increase profitability of the operation assuming no other changes.

Sensitivity Analysis for Net Income per Sow

A sensitivity analysis available in the model can help determine how changing various assumptions impacts net income. The model considers differences in weaned pig price, cull sow value, and replacement gilt cost across a range of sow or pre-wean mortality rates as shown in Table 4. It is important to note that only the two variables in each sensitivity table change and all other production, price and cost inputs and assumptions are held constant at their original values used in the current scenario. The budget comparisons of the current and improved scenario may not match the net income per sow in the sensitivity tables if these other inputs are changed in the improved scenario. For the first three tables showing a range of total sow mortality, the percent sow mortality from farrowing to breeding and pre-wean mortality are 7% and 15%, respectively. For the fourth table, with farrowing-to-breed mortality, the total sow mortality is 14%.

The first three sensitivity tables consider a change in total sow mortality and changes in weaned pig price, cull sow price, and replacement gilt price. The fourth shows the effect of changing sow mortalities that occur from farrowing-to-breeding and weaned pig price. This changes the litters and number of pigs born but does not change cull sow income or replacement gilt cost, which is included in total sow mortality. Pre-wean mortality is also an important factor affecting profitability in breed-to-wean operations. A sensitivity analysis using the same inputs as the current scenario shows that a one percentage point change in pre-wean mortality, i.e., 15% to 14%, affects net income per sow by $17.23, in this example with weaned pig price at $58 per head. This net income effect increases as weaned pig price increases and vice versa.

Table 4. Sensitivity tables for annual net income per sow

Table 4. Sensitivity tables for annual net income per sow

		Weaned pig price				
Net income per sow		**$54.00**	**$56.00**	**$58.00**	**$60.00**	**$62.00**
	12.00%	$483.87	$535.00	$586.13	$637.27	$688.40
Percent total sow mortality	**13.00%**	$469.27	$520.09	$570.90	$621.71	$672.52
	14.00%	$454.68	$505.17	$555.66	$606.15	$656.64
	15.00%	$440.09	$490.25	$540.42	$590.59	$640.76
	16.00%	$425.49	$475.34	$525.19	$575.03	$624.88

		Cull sow price				
Net income per sow		**$44.00**	**$46.00**	**$48.00**	**$50.00**	**$52.00**
	12.00%	$577.67	$581.90	$586.13	$590.36	$594.59
	13.00%	$562.62	$566.76	$570.90	$575.04	$579.18
Percent total sow mortality	**14.00%**	$547.56	$551.61	$555.66	$559.71	$563.76
	15.00%	$532.50	$536.46	$540.42	$544.38	$548.34
	16.00%	$517.45	$521.32	$525.19	$529.06	$532.93

		Replacement price				
Net income per sow		**$355.00**	**$365.00**	**$375.00**	**$385.00**	**$395.00**
	12.00%	$597.53	$591.83	$586.13	$580.43	$574.73
	13.00%	$582.50	$576.70	$570.90	$565.10	$559.30
Percent total sow mortality	**14.00%**	$567.46	$561.56	$555.66	$549.76	$543.86
	15.00%	$552.42	$546.42	$540.42	$534.42	$528.42
	16.00%	$537.39	$531.29	$525.19	$519.09	$512.99

		Weaned pig price				
Net income per sow		**$54.00**	**$56.00**	**$58.00**	**$60.00**	**$62.00**
	5.00%	$437.31	$487.16	$537.01	$586.85	$636.70
Percent sow mortality farrowing-to-breeding	**6.00%**	$446.00	$496.16	$546.33	$596.50	$646.67
	7.00%	$454.68	$505.17	$555.66	$606.15	$656.64
	8.00%	$463.36	$514.18	$564.99	$615.80	$666.61
	9.00%	$472.05	$523.18	$574.31	$625.45	$676.58

*Total mortality is 14%.

		Weaned pig price				
Net income per sow		**$54.00**	**$56.00**	**$58.00**	**$60.00**	**$62.00**
	13.00%	$486.76	$538.43	$590.11	$641.79	$693.47
	14.00%	$470.72	$521.80	$572.89	$623.97	$675.05
Pre-wean mortality	**15.00%**	$454.68	$505.17	$555.66	$606.15	$656.64
	16.00%	$438.64	$488.54	$538.43	$588.33	$638.23
	17.00%	$422.60	$471.91	$521.21	$570.51	$619.81

The relative effect of a change in mortality rates, weaned pig price, cull sow price, and replacement gilt price in this example is summarized in Table 5.

Table 5. Relative effect of a change.

Table 5. Relative effect of a change

Relative change	Net income per sow change
One percentage point in total sow mortality	$14.59 to $15.88
$2.00 per head in weaned pig price	$49.85 to $51.13
$2.00 per cwt in cull sow price	$3.87 to $4.23
$10 per head in replacement gilt cost	$5.70 to $6.10
One percentage point change in sow mortality from farrowing-to-breeding	$8.68 to $9.97
One percentage point change in pre-wean mortality	$16.04 to $18.41

Summary

Sow mortality and pre-wean mortality rates affect productivity and potential profit in breed-to-wean swine production. Assessing the economic opportunity of improving mortality rates is important when considering approaches and costs to accomplish pig survivability goals.

The assessment presented here provides an example of possible net income changes due to a reduction in mortality for a representative operation. It is not intended to represent any one operation, production system, or firm, which would be defined by notable variation in specific situations and managerial approaches. Users are encouraged to use this and related resources to conduct their own economic analyses with the coordinating Ag Decision Maker Decision Tool spreadsheet.

Read More: https://porkgateway.org/resource/economic-assessment-of-mortality-in-wean-to-finish-production

References and Additional Resources

[1] PigCHAMP Benchmarking summaries

[2] When are sows dying? Is it different in the US compared to Canada, Ron Ketchem, Mark Rix, Valerie Duttlinger, National Hog Farmer Feb 28, 2018. Tracking Sow Mortality, Tom Gillespie, PIC Benchmark 2018 Spring Edition. Sow mortality in the US: An industry-wide perspective, John Deen DVM PhD, Jinliang Xue DVM PhD. University of Minnesota-Twin Cities, 1999 Leman Swine Conference proceedings.

This project was supported by the National Pork Board and the Foundation for Food and Agriculture Research grant #18-147.

Elanco

Farrowing House Management Effect on Total System Productivity

John Sonderman, Brady McNeil and Amanda Cross

DNA Genetics, 4438 Old Mill Court, Columbus, NE 68601, USA
Email: jsonderman@dnaswinegenetics.com

1. Introduction

Improvements in total born and born alive have changed the paradigm on how to best maximize piglet throughput in a farrowing house. The US swine industry averaged 10.3 born alive in 2004, compared to 13.9 born alive in 2022 (PigCHAMP, 2004; PigCHAMP, 2022). According to PigChamp (Q2 2022) the top 10% of US producers are averaging 15 born alive. This increase in number born alive has producers evaluating different sow loading strategies and incorporating nurse sows into standard practice. These strategies may work for some producers but are they the correct approach when evaluating profitability on a total systems approach?

2. Loading Female Strategies

There are various strategies that farrowing barn managers implement in the first 24 hours to maximize the number and quality of pigs out the door. Strategies have to do with teat count, parity, and history of the females in the farrowing room as well as the health status of the barn.

When determining how to load sows in a farrowing room, limiting pig movements as much as possible to maximize the number of females with her own pigs only (HOPS) is key. Pigs moved to different litters, as well as the pig in the litter they were moved to, had lower growth rates than pigs in litters without a disturbance (Robert and Martineau, 2001). McCaw (2000) looked at two different movement strategies in a PRRS positive farm. Size sorting was the first movement strategy, which resulted in 85% of the pigs being moved off their birth dam (McCaw, 2000). Moving pigs based on litter size within 24 hours was the second movement strategy, which resulted in 15% of the pigs being moved off their birth dam (McCaw, 2000). Size sorting resulted in 14.9% pre-wean mortality compared with 9.8% pre-wean mortality in the low movement group (McCaw, 2000). The timing of cross-fostering is important. Cross-fostering too soon may result in the pigs not receiving colostrum. However, cross-fostering too late may result in decreased acceptance by the cross-foster sow or decreased integration into the new litter due to establishment of teat order (Heim, 2012). The sow's lactation success could be impacted by the timing of cross-fostering. The number of pigs suckling at 24 hours was the most important factor in transitory milk yield or the first milk after colostrum (30-61 hours post farrow, Vadmand 2015).

Teat count is a trait farms can use when identifying which females to bring into the herd. Multiplier farms will count total teats at birth as an easy way to identify if a female makes the minimum qualification for a replacement gilt. Each teat added to a female is worth an extra 0.34 pigs weaned and as teat count increased, individual pig intake and total sow colostrum production increased (Wiegert and Knauer, 2018).

As number born alive has increased, so has the day-to-day variation in born alive. This presents management challenges for farrowing leads. Considering teat count is one way to help manage increased variation in viable pigs. One Pillen Family Farms sow farm has an average daily born alive that ranges between 13.5 and 15.5 pigs (Figure 1).

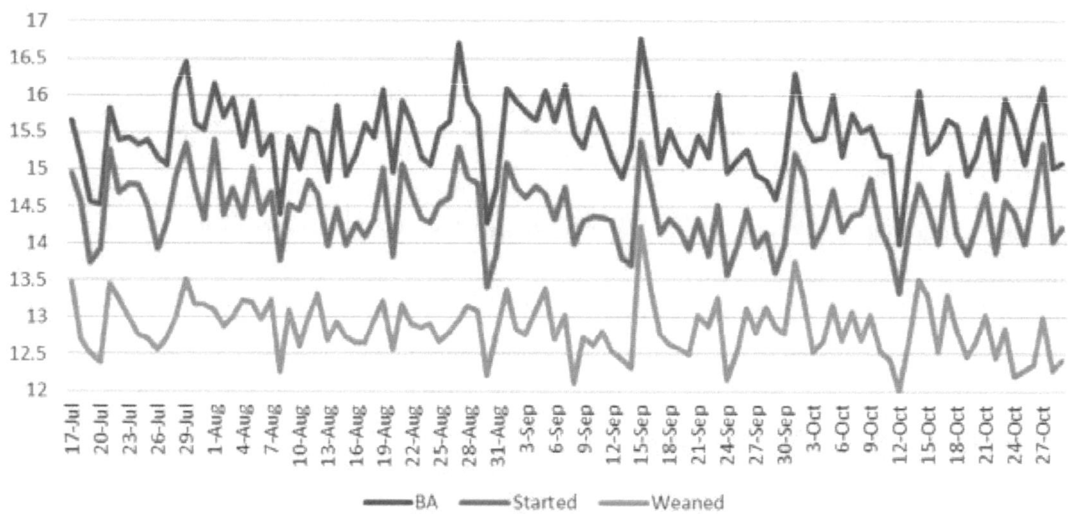

Figure 1. PFF farm in McRebel situation comparing number Born Alive vs. Number Started vs. Number Weaned by farrow day

As the farrowing managers determine where to foster the excess pigs, they should consider teat count of the sows. When increasing number started from 14 to 15, there is an increase in pig survival when the gilt has 16 teats (79%) compared to 14 teats (34%) (Table 1). Sows display a similar trend in increased pig survival as teat count increased (Table 2). Farrowing managers should also consider parity when managing variation in viable pigs. Average number weaned from DNA females of all health statuses is 11.7 for parity 1 sows and 10.6 for parity 7 and older sows (SMS, 2022). As parity increases, the number weaned increases to parity 5. Pigs placed on parity 3 to parity 5 females, regardless of their birthweight, had higher weaning weights and are more likely to be consumers of feed upon entry into the nursery (Huting et al., 2019).

Table 1. Parity 1 additional pigs weaned for each additional pig started based on teat count

	P1	Teat Count			
Started		14	15	16	AVG
	11	0.91	1.33	1.54	1.09
	12	0.82	0.75	0.63	0.78
	13	1.06	1.24	0.57	1.06
	14	0.67	0.71	1.14	0.75
	15	0.34	0.46	0.79	0.42
	16	0.35	1.05	0.53	0.58
	17	0.34	-0.32		0.14
	18	0.51			

Table 2. Parity 2+ Additional Pigs Weaned for each additional pig started based on teat count

	Parity 2+ Teat Count				
Number Started	13	14	15	16	AVG
11		1.19	0.64	1.19	0.94
12	1.09	0.68	0.88	0.81	0.78
13	0.24	0.79	0.82	0.48	0.73
14	0.78	0.81	0.79	1.01	0.83
15	0.17	0.33	0.60	0.58	0.43
16	-0.12	0.29	0.52	0.42	0.36
17	0.95	0.75	0.44	0.58	0.63
18		0.02	0.23	0.57	0.16
19		-0.02	0.01		

Lastly, health status of the barn should be considered when determining the overarching loading strategy of sows. Many farms in a disease closure use a McREBEL approach to reduce exposure to bacteria to eliminate losses. This McREBEL process eliminates any movement of pigs between crates. Therefore, no pigs are moved, and all sows wean her own pigs.

3. Nurse/Fall Behind Sow Strategies

Many farms will utilize a fall behind or nurse sow for piglets that are not thriving in their current litter. There are many variations in nurse sow strategies, but typically pigs are moved to a nurse sow between day 3 and 5 post farrow. One option is a 1 step nurse sow, where a wean sow is brought back to a 3 to 5-day old room. This sow will then have fall behind pigs placed on her or she will replace a sow in that room that is milking well, and that sow is used for fall behind pigs. Another option is a 2-step nurse sow, where a weaned sow is moved to a day 10 sow and a day 10 sow is brought to a day 3 to 5 room.

When moving pigs to a nurse sow, team members go through the farrowing room on day 3 to 5 and move the lightest pigs. In an internal Pillen Family Farms trial, the impact of nurse sows was investigated. Farrowing rooms were assigned to two different treatments, either create a nurse sow (Treatment = Yes) or no nurse sow (Treatment = No). Light pigs in the non-nurse sow treatment room were identified as fall behind pigs, but the farrowing team was not allowed to move these pigs to a nurse sow. Pigs that were in the non-nurse sow treatment room and marked to be moved to a nurse sow (Fall Behind Status = 2, Figure 2) or pigs that were moved to a nurse sow (Fall Behind Status = 1, Figure 2), were 0.63 to 0.86 pounds lighter than those pigs that stayed on their original dam (Fall Behind Status = 3, Figure 2). Birth weights between pigs that had fall behind pigs marked or removed and remained on their original dam (Fall Behind Status = 3, Figure 2) or pigs that did not have any pigs in the litter marked for nurse sows (Fall Behind Status = 4, Figure 2) were very similar. Pigs that were moved to nurse sows (Fall Behind Status = 2, Figure 3) were 2.2 pounds heavier at weaning than pigs that were marked as fall behind pigs and not moved to a nurse sow (Fall Behind Status = 1, Figure 3). Also, non-fall behind pigs that remained in the litter with the fall behind pigs (Treatment = No, Fall Behind Status = 3, Figure 3) gained less at weaning than the pigs that had fall behind pigs removed from their litters (Treatment = Yes, Fall Behind Status = 3, Figure 3).

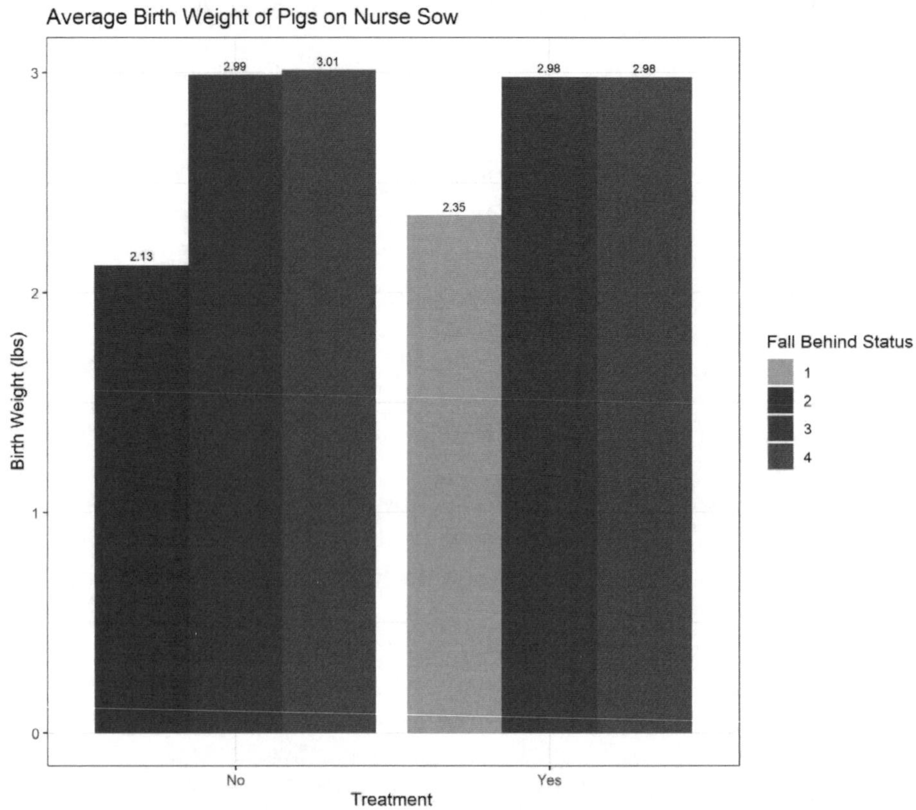

Figure 2. Average birth weight of pigs moved to nurse sows and those that remained on their original dams.

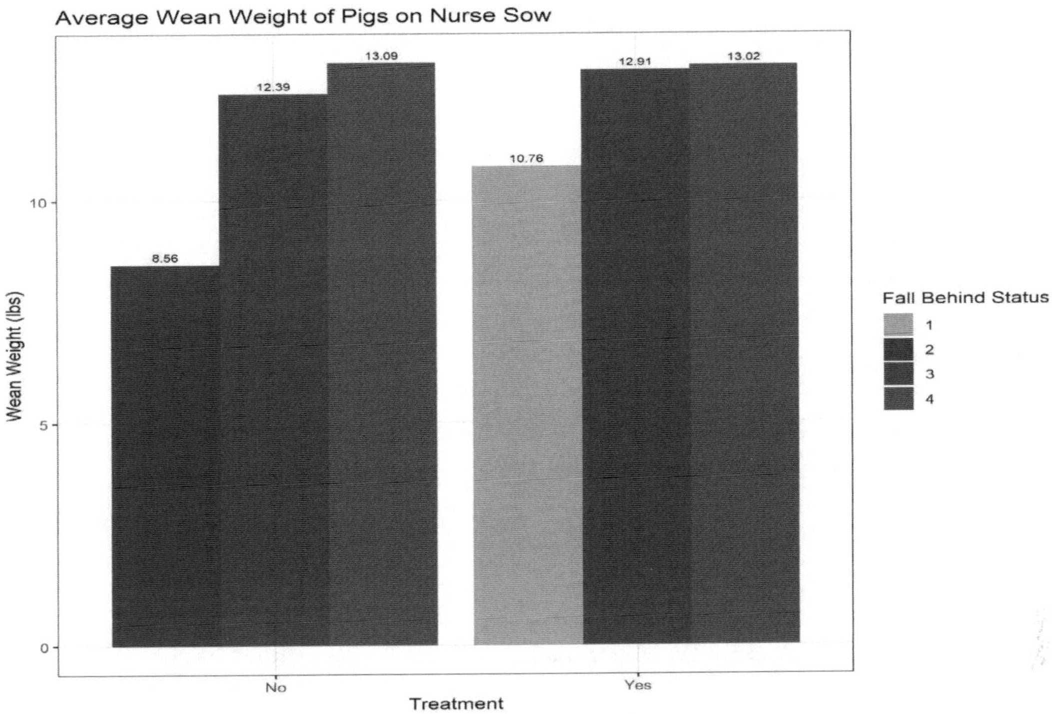

Figure 3. Average wean weight of pigs moved to nurse sows and those that remained on their original dams.

Another internal Pillen Family Farms trial was conducted to compare two similar farms under different management processes. The first farm was at the end of a PRRS closure and performing strict McREBEL (Figure 1), the second farm was PRRS negative, moving pigs at different time points, and utilizing 15% nurse sows. While the farm utilizing 15% nurse sows weaned more pigs on average across all parities, the strict McREBEL farm weaned more in the key parity range of 2-5 (Table 3).

Table 3. Pigs weaned by parity in a McRebel vs. high piglet movement scenario.

Parity	Average Weaned		
	No Movement	15% Nurse Sows	Difference
1	11.63	12.28	-0.65
2	12.61	11.88	0.73
3	12.45	11.41	1.04
4	12.23	11.23	1
5	12.05	11.33	0.72
6	11.47	11.37	0.1
7	11.33	11.44	-0.11
8	10.46	11.29	-0.83
9	11.09	13	-1.91
10	10.23		
Total	11.66	12.01	

4. Impact of Farrowing House Management on Post Wean Performance

Post weaning performance of pigs should also be considered when evaluating sow management strategies. Providing pigs with the correct environment early in life can set them up for lifetime success. McCaw (2000) reported a 5% difference in nursery survivability in the face of a PRRS challenge based on farrowing house movement. Internal Pillen system data comparing a PRRS positive flow with no piglet movement (McREBEL) and a PRRS negative flow with 15% nurse sows, showed that the pigs from the farm with no piglet movement had 1.14% less nursery mortality (1.37 vs. 2.51) while having a younger wean age (21 vs. 25).

5. Conclusion

Each day farm managers and farrowing leads are challenged with how to produce the most amount of pigs for the least cost. The process to produce the most pigs may vary based upon factors specific to that farm and to that day. Overall, however, loading up the correct females based on parity and teat count and then leaving litters intact will best set the farm and total system up for weaning and post weaning success.

References

Heim G., Mellagi A.P.G., Bierhals T., de Souza L.P., de Fries H.C.C., Piuco P., Seidel E., Bernardi M.L., Wentz I., Bortolozzo F.P., Effects of cross-fostering within 24 h after birth on pre-weaning behaviour, growth performance and survival rate of biological and adopted piglets. Livest. Sci. 2012; 150:121–127. doi: 10.1016/j.livsci.2012.08.011.

Huting A.M.S., Sakkas P. and Kyriazakis I., Sows in mid parity are best foster mothers for the pre- and post-weaning performance of both light and heavy piglets, *Journal of Animal Science*, Volume 97, Issue 4, April 2019, Pages 1656–1670, https://doi.org/10.1093/jas/skz062

McCaw M.B., Effect of reducing crossfostering at birth on piglet mortality and performance during an acute outbreak of porcine reproductive and respiratory syndrome. *Swine Health Prod.* 2000;8(1):15-21.

PigCHAMP. (2004). PigCHAMP Benchmarking USA year summary. https://www.pigchamp.com/benchmarking/benchmarking-summaries.

PigCHAMP. (2022). PigCHAMP Benchmarking USA quarterly summary. https://www.pigchamp.com/benchmarking/benchmarking-summaries.

Robert S. and Martineau G. P., Effects of repeated cross-fosterings on preweaning behavior and growth performance of piglets and on maternal behavior of sows, *Journal of Animal Science*, Volume 79, Issue 1, January 2001, Pages 88–93, https://doi.org/10.2527/2001.79188x

Vadmand C. N., Krogh U., Hansen C. F., Theil P. K., Impact of sow and litter characteristics on colostrum yield, time for onset of lactation, and milk yield of sows, *Journal of Animal Science*, Volume 93, Issue 5, May 2015, Pages 2488–2500, https://doi.org/10.2527/jas.2014-8659

Wiegert J.G. and Knauer M.T., 98 Sow Functional Teat Number Impacts Colostrum Intake and Piglet Throughput., *Journal of Animal Science*, Volume 96, Issue suppl_2, April 2018, Pages 51–52, https://doi.org/10.1093/jas/sky073.096

Grand MANUFACTURING CO. LTD.

Stainless Steel Hog Feeders

"The Industry's Newest Leading Wet/Dry Feeding Technology"

Grower / Finisher Feeder

- Comes in 30", 45", 60", 75", & 90" lengths
- 30" has two feed spaces per side, 15 animals per feed space
- 45" has three spaces per side, 15 animals per feed space
- Feeders designs for management simplicity.

Weanling Feeder

- Comes in 30", 40", 50", & 60" lengths
- 30" has three feed spaces per side, 10 animals per feed space
- 40" has four spaces per side, 10 animals per feed space
- Handles crushed feed or pallets with high moisture or high fat feeds.

Extension Hopper
(Optional add-on for all feeder sizes)

Adjuster
(Basic incremental feed adjuster)

Grand MANUFACTURING CO. LTD.

Box 10 Oakville, Manitoba R0H 0Y0
ph: (204) 267.2196 fax: (204) 267.2458
garry@grandmfg.com www.grandmfg.com

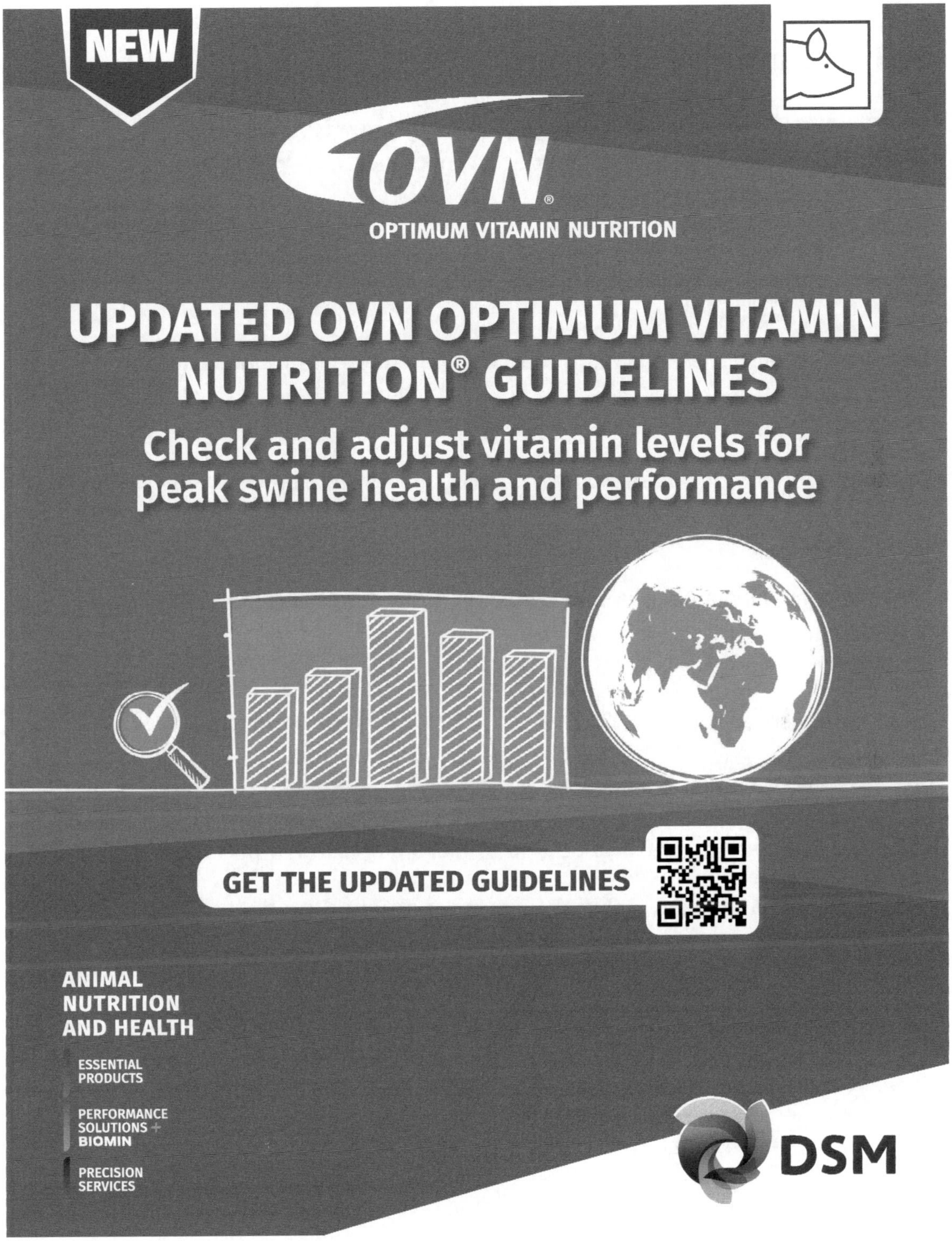
NEW
OVN®
OPTIMUM VITAMIN NUTRITION
UPDATED OVN OPTIMUM VITAMIN NUTRITION® GUIDELINES
Check and adjust vitamin levels for peak swine health and performance
GET THE UPDATED GUIDELINES
ANIMAL NUTRITION AND HEALTH
ESSENTIAL PRODUCTS
PERFORMANCE SOLUTIONS + BIOMIN
PRECISION SERVICES
DSM

Strategies to Maximize Profit in Late Finishing

Matthew Ritter[1], Christian Bruneau[2], Karl De Ridder[2], Molly McGhee[1], and Zhenbin Zhang[2]

[1]Cargill Animal Nutrition, 6571 State Route 503, Lewisburg, OH 45338
[2]Cargill Animal Nutrition, 404 Main St., Woodstock ON N4S1T5
Email: matthew_ritter@cargill.com

1. Introduction

The objective of this presentation is to outline strategies in late finishing and pig marketing that can be used to maximize producer profitability. Futures markets, economic models, and research trials will be used to demonstrate how optimal market weight, diet strategy, finishing space / time, and managing variation in market weight will play a role in maximizing profit during the summer of 2023.

2. Study Market Conditions and Set System Goals

Prior to placing weaned pigs, it is important to study futures markets for feed ingredient costs and carcass price and to understand the implications this has on system goals, profitability, optimal market weight, and production constraints. In the U.S., future prices for feed ingredients (corn and soybean meal) and lean hogs are traded by the Chicago Board of Trade and the Chicago Mercantile Exchange, respectively. This information can then be used as inputs in profitability models like Cargill's Pig Flash Model to forecast total cost, total revenue, and net profit on a $/pig basis for the next 12 months. Canadian producers need to make adjustments for Canada – U.S. exchange rates, carcass dressing percentage, and carcass index – to customize this information for their markets (e.g., Ontario Pork's 100% Formula). The key point of this exercise is to understand if weaned pigs placed in January 2023 will be profitable when marketed in June 2023 and how this will impact management decisions for the next 6 months.

Once forecasted feed costs and carcass price estimates are known for June 2023 marketings, we can work through an example of calculating optimal market weight for one packer. Optimal market weight models like Cargill's Producer Profitability require the following inputs: feed cost, finishing growth curve [(average daily gain (ADG), average daily feed intake (ADFI), and feed efficiency (FE)], carcass price, packer payment grid, and individual carcass data. Pork packer grids in Canada vary from packer to packer but are structured in a manner that offer premiums / discounts based on carcass weight and carcass quality measurements (e.g., carcass yield, backfat, loin depth, and/or carcass index categories). Optimal market weight can be calculated in two steps: 1) single pig optimum that calculates net profit on 1 kg intervals; and 2) population optimum, which takes into account market weight variation. The following assumptions were made at the time of paper submission: 1) wean to market feed cost per kg of gain = $1.43; 2) last finishing diet cost = $450/tonne; 3) carcass price = $2.56/kg; and 4) market weight coefficient of variation = 10% (Table 1). Figure 1 plots net profit by live weight and demonstrates that: 1) net profit is maximized at 156 kg live weight; and 2) the live weight range of 131-156 kg is within $10/pig net profit of the single pig optimum. However, producers need to be aware of the large discounts placed on pigs marketed >156 kg live weight in this scenario, and thus market weight variation must be considered.

Table 1. Economic assumptions for June 2023 marketings

Model Inputs	Values
Start weight, kg	6.5
End weight, kg	136
Feed cost per kg gain, $/kg	$1.43
Last diet cost, $/tonne	$450
Yardage and labor, $/day	$0.155
Carcass price, $/kg	$2.56
Market weight CV, %	10%

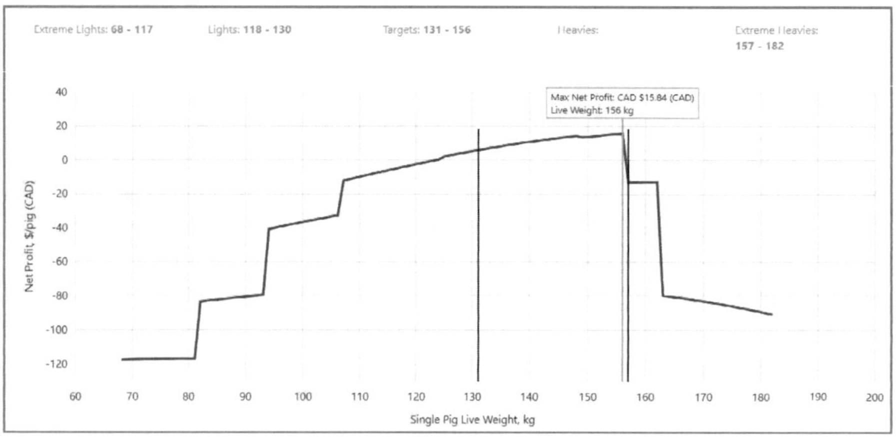

Figure 1. Net profit by live weight (single pig optimum)[1, 2, 3]
[1] Based on assumptions in Table 1 for June 2023 marketings
[2] Single pig optimum where net profit is maximized = 156 kg
[3] Target weight range is defined as any live weight that has a net profit within $10 / pig of the single pig optimum ($15.84/pig)

Figure 2 then looks at the plot of net profit vs. population average weight using the current distribution of marketed carcasses [n = 5,000 pigs; mean = 129 kg; live weight coefficient of variation (CV) = 10%]. This approach takes into account that as the average weight drops, a greater percentage of pigs will be discounted for being too light, and on the flip side, as average weight increases, a greater percentage of pigs will be discounted for being too heavy. The population average weight where net profit is maximized in this scenario is 135 kg live weight (net profit = $3.89/pig).

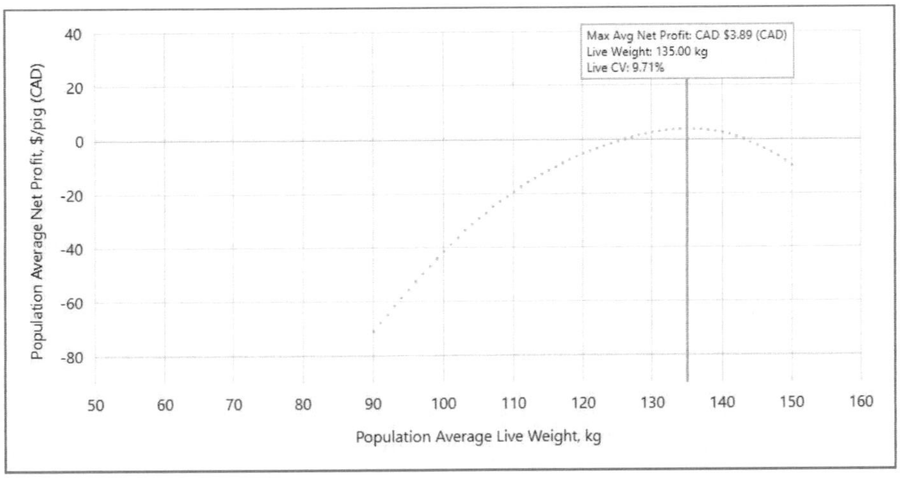

Figure 2. Net Profit by average live weight (population optimum)[1,2,3,4]
Based on assumptions in Table 1 for June 2023 marketing
[2] Based on the net profit by live weight values in Figure 1
[3] Based on current marketings in live weight (n = 5,000 pigs; mean = 129 kg; live weight CV = 10%)
[4] Population optimum where net profit is maximized = 135 kg

3. Late Finishing Profit Strategies

Now that we have projected the population optimum weight to be 135 kg live weight for June 2023, we need to understand current marketing, historical performance, and system constraints. In this example, the producer's current live market weight is 6 kg lower than the optimum weight, and this comes at a cost of $2.03/pig (Table 2). Below is a list of management strategies that can be used to close the gap between current and optimal market weights for our working example above.

Table 2. Current marketings vs. population optimum[1]

Metric	Current Marketings	Population Optimum	Difference
AVG Live WT, kg	129	135	6
Live CV, %	9.71	9.71	0.00
Extreme Lights, %	20.06	9.08	-10.98
Total Lights, %	56.66	36.57	-20.09
Target, %	42.12	58.41	16.29
Total Heavies, %	1.22	5.02	3.80
Extreme Heavies, %	1.22	5.02	3.80
Net Profit, $/pig (CAD)	1.85	3.89	2.03

[1]Based on population optimum modeling from Figure 2

3.1 Diet Strategy and Feed Additives

Diet strategy and feed additives will be covered in detail by Kurt Stoess of HyLife. For the purposes of this discussion, it's important to model both performance and economic projections for various levels of dietary energy (added fat), lysine, by-products (e.g., DDGS), and feed additives to: 1) identify the most profitable feeding strategies; and 2) determine if the optimal weight of 135 kg can be achieved with the system's current finishing space.

3.2 Additional Finishing Space

Given market restrictions on the use of ractopamine and the high price of adding fat to diets, many producers are exploring the option of adding additional finishing space, so their pigs have more time to reach the optimal market weight. In the current example (Table 1), we can calculate the net return of adding finishing space as means to increase live market weight from 129 to 135 kg. The expected performance values from the growth curve for this weight range are: ADG = 0.85 kg/day and FE = 3.68. The 6 kg increase in body weight requires 7 more days

($0.155/day) and 22 kg more feed ($450/tonne), resulting in ~$11/pig in additional cost. Meanwhile, the 6 kg of additional live weight (4.8 kg carcass weight and carcass price = $2.56/kg) represents $12.29/pig in additional revenue. The net return for 1 additional week of finishing space in this scenario is $1.29/pig, so this becomes a viable option to explore for the summer of 2023.

3.3 Managing Variation in Market Weight

Kansas State University and Cargill collaborated on a meta-analysis of wean-to-finish trials with 117,268 individually weighed pigs to predict live weight variation within a group of pigs. This work suggests that the expected CV in live weight for grow-finish pigs weighing 100-130 kg is ~10% (Tolosa et al., 2021). Please note that each additional week of barn fill-time increases CV by ~0.6% (Ritter et al., 2021) and health challenges like PRRS can have a major impact on CV (Cornelison et al., 2018). Managing this variation in market weight then has huge implications for producer profitability as reducing market weight CV from 10% to 9% has been estimated to be worth between $0.75 to $1.75 per pig (Ritter et al., 2021).

Optimal marketing strategies need to balance production constraints (space and packer commitments), late finishing performance, current vs. optimal market weights, and packer payment grids. Ritter et al. (2022) recently evaluated the effects of four different marketing strategies on average market weight and growth performance traits in heavy weight pigs (n = 2,748) over the last 28 days of finishing: 1) whole pen dump (100% marketed on day 28); 2) cut and pen dump

(25% on day 0 and 75% on day 28); 3) two cuts and pen dump (25% on day 0, 25% on day 14, and 50% on day 28); and 4) five equal weekly cuts (20% on day 0, 7, 14, 21, and 28). The results of this trial demonstrate that marketing strategy has a major impact on market weight, ADG, and FE (Table 3). Taking at least one cut (Trt 2, 3, and 4) improved ADG by ~20% and FE by ~10% over the 28-day marketing period. However, decreasing the total number of cuts from 5 (Trt 4) to 2 (Trt 2), increased live weight gain by 7 kg per pig, while having with no change in FE. Therefore, producers who are short on space and need to add weight can utilize a cut and pen dump marketing strategy to increase days and market weight without compromising late finishing FE.

Table 3. Effects of marketing strategies on late finishing growth performance traits[1]

Measurements	Trt 1 Whole Pen Dump	Trt 2 Cut and Pen Dump	Trt 3 Two Cuts and Pen Dump	Trt 4 Five Equal Weekly Cuts
Pens	22	22	22	22
Pig marketed, %				
Day 0	---	25%	25%	20%
Day 7	---	---	---	20%
Day 14	---	---	25%	20%
Day 21	---	---	---	20%
Day 28	100%	75%	50%	20%
Average days on feed	28	21	17.5	14
Start weight, kg	120.8	118.8	120.1	119.9
Market weight, kg	142.2[a]	138.8[ab]	135.8[bc]	133.1[c]
Average weight gain, kg/pig	21.3[a]	19.9[a]	15.5[b]	12.9[c]
ADG, kg/day	0.75[a]	0.94[b]	0.89[b]	0.85[ab]
ADFI, kg/day	2.56[a]	2.89[b]	2.74[ab]	2.63[ab]
Feed:Gain	3.38[a]	3.03[b]	3.02[b]	3.02[b]

[1] Adapted from Ritter et al. (2022)

[a,b,c] Means with different superscripts differ ($P < 0.05$)

Another strategy to consider for adding weight and reducing light weight pigs is to bump standardized ileal digestible (SID) lysine after the first cut has been marketed from a barn. A recent Cargill trial demonstrated increasing SID lysine/net energy ratio from 2.85 to 2.99 for 21 days after the first cut (25%) was marketed (body weight = 131 kg) improved ADG by 9%, FE by 5%, and final body weight by 0.7 kg for pigs remaining in the pen. It is important to note that in this trial, the population average live weight at the timing of the first cut was ~120 kg and mixed gender pens were utilized. The population of pigs remaining in the pen after the first cut was taken averaged ~116 kg and was skewed heavier towards gilts vs. barrows, so this gives some insight as to why late finishing pigs may respond to a bump in SID lysine after the first cut.

4. Summary

Future markets are forecasting profits for the summer of 2023. The plan for maximizing summer profits starts today with modeling optimal market weight and diet strategy to determine if finishing space, marketing strategies, and late finishing diets will need to be adjusted to meet your goals.

References

Tolosa, A. F., J. M. DeRouchey, M. D. Tokach, R. D. Goodband, J. C. Woodworth, J. T. Gebhardt, M. J. Ritter, and C. M. Pilcher. 2021. A meta-analysis to understand the relationship between pig body weight and variation from birth to market. *Animals. 11*:2088.

Cornelison, A. S., L. A. Karriker, N. H. Williams, B. J. Haberl, K. J. Stalder, L. L. Schulz, and J. F. Patience. 2018. Impact of health challenges on pig growth performance, carcass characteristics, and net returns under commercial conditions. Transl. Anim. Sci. 2(1):50-61.

Ritter, M., A. Tolosa, and J. DeRouchey. 2021. Roundtable hosted by Laura Greiner:

Why should we care about pig body weight variation? Swine IT Podcast. July 19, 2021.

Ritter, M., S. Ebarb, C. Hagen, C. Parks, J. Erickson, B. Ramer, and B. Frederick. 2022. Effects of market strategies on growth performance traits of heavy weight finishing pigs. Proceedings of the Allen D. Leman Swine Conference, St. Paul, MN.

Cargill®
Don't miss
the chance to
WIN BIG!
Scan the QR code and complete the form!
Pour accéder au formulaire en français
4 BIG WINNERS
Cargill Pork Thrive Challenge 2023
Travel to World Pork Expo, Iowa/US – June, 2024
(Hotel + transportation + registration + meals)
$1,500 to invest on your swine business
+
On-farm consultation from a Cargill expert
Cargill/Purina Feed Credit ($1,500)
+
Access to Cargill's farm management tools tailored to your farm needs for one year
NHL Hockey Game*
(Pair of tickets)
+
Dinner
+
Transportation
*Distance limitation will be applied

Wean-Finish Efficiencies - The Often-Overlooked Areas

Nat Stas

Pig Improvement Company, 100 Bluegrass Commons Blvd, Ste 2200, Hendersonville, TN 37075
Email: nat.stas@genusplc.com

1. Introduction

Initially when we think about wean-to-finish efficiency, our minds go to diet and feed management which is logical since dietary energy can account for 50%-60% of the cost of production and minimizing feed waste can have a 5%+ impact on feed conversion. We will touch on feed and space management due to this, however, there are many other areas to consider that impact production efficiency. In addition to the common areas, we will review the impact of wean weight efficiency, leading to a quality weaned pig, the importance of monitoring and staying on the feed budget, what affect mortality can have on efficiency, and the thermal comfort of the pig, an important topic with fuel energy costs lately. A practical approach is the goal as we always need to consider our labor efficiencies.

2. The Common Areas

2.1 Feed Delivery, Feeder Space and Feeder Adjustment

First and foremost, feed needs to be delivered in a timely manner and needs to flow to the feeder without spillage. It is estimated that 2-10% of feed that is delivered is wasted, either at the bin, the boot, the feed line, or the feeder. Feeder space is always a consideration because it can greatly impact feeding behavior and how we should manage feeder adjustment. PIC recommends 2.5cm per pig on a dry feeder in the nursery, 5cm per pig on a dry feeder in finishing and 3.2cm per pig on a wet/dry feeder in finishing. If feeder space is long or short, it can be often be compensated with floor space (Figure 1.) and/or pan coverage.

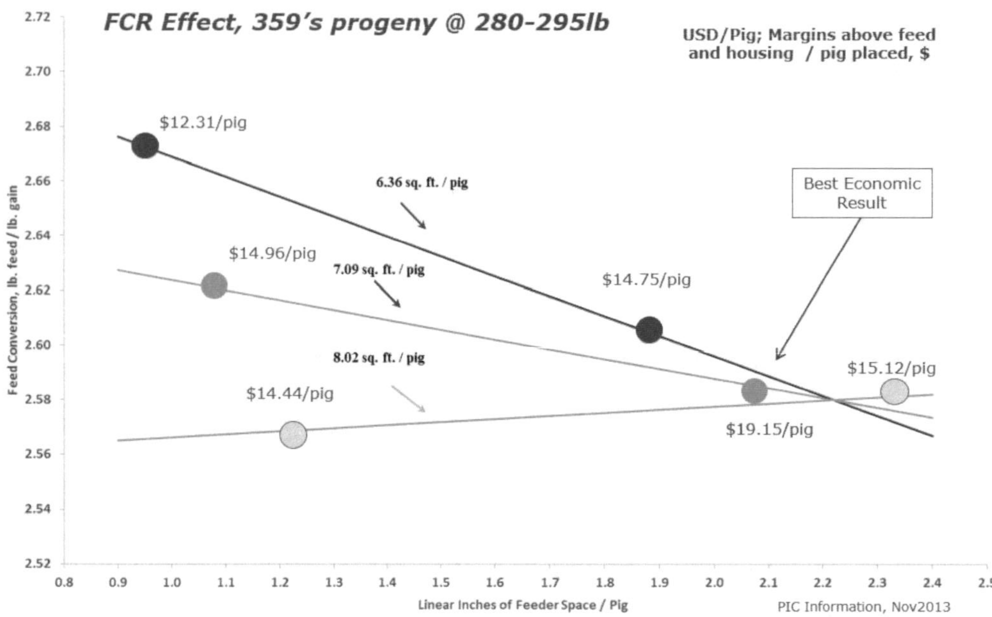

Figure 1. Feeder space vs feed conversion.

Excessive feeder space can lead to significant feed wastage if not properly managed; visualize feeder pan coverage in hospital pens and after first cuts if not adjusted. Last but certainly not least, feeder adjustment should target ~50% pan coverage. Feeder adjustment that is too tight, leading to <25% pan coverage can impact FCR by ~1.5% and adjustment that is too loose, leading to >75% pan coverage can impact FCR by ~2.5%.

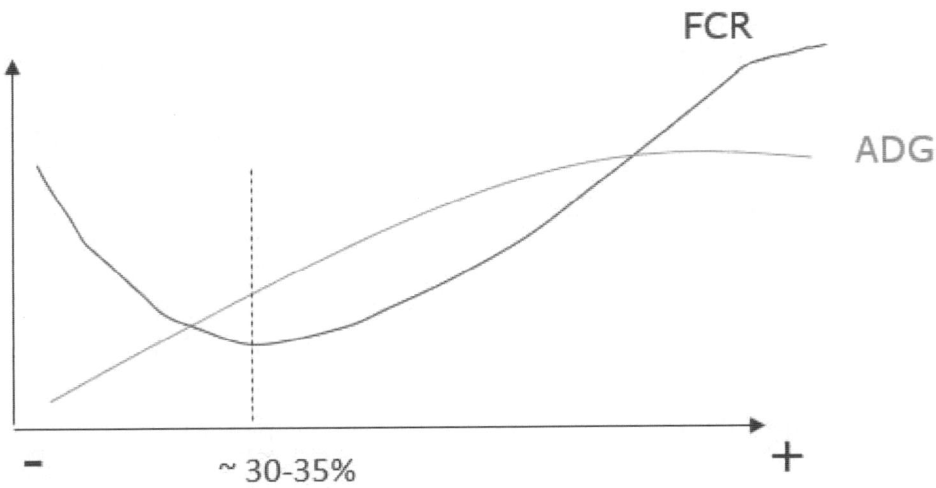

Figure 2.FCR & ADG

2.2 Stocking Density

As briefly discussed above, feed space and floor space work together to drive efficiencies in pig production. In conjunction with the feeder space recommendations above, PIC recommends 0.67m^2 per pig. This is based on current market weights and Gonyou's et al. (2006) formula keeping the *k-value* at or above 0.0336. However, to drive better throughput and efficient use of space within a barn, a multi-cut marketing strategy should be implemented. Lerner et al. (2020) showed starting stocking density at 0.66m^2 per pig and implementing 4 marketing events starting at d 93, d 122, and d 147, ending stocking density at 0.71m^2 resulted in similar BW, ADG, and FCR when compared to a stocking density of 0.71m^2 from the start.

2.3 Out-Of-Feed Events (OOFE)

It has been shown that out-of-feed events decrease pig efficiency without recovery. Prairie Feed Systems and JBS United showed a decrease of 1% and 2.3% in FCR for a 24hr OOFE once per week and once every other week, respectively. It is also important to note an OOFE can occur on a smaller scale and impact the eating order, resulting in increased market variation, overall decreasing pounds sold per barn and impacting barn efficiency.

3. The Over-Looked Areas

3.1 Wean Weight Efficiency: Wean Weight and Wean Age

It is important to understand what the potential of the pigs we are raising is. Starting with genetic potential, the pigs are capped at what they can perform. Secondary to genetics, the pigs rearing environment can have a great impact on downstream wean-to-finish performance. This should be measured by means of wean weight efficiency, meaning the pigs actual wean weight per day of age versus what is to be expected.

Table 1. Expected wean weight

Age	Weight, lbs	Age	Weight ,lbs
18	11.7	**24**	15.4
19	12.1	**25**	16.3
20	12.8	**26**	16.9
21	13.4	**27**	17.8
22	14.1	**28**	18.5
23	14.7		

Based on the table above, we can calculate if a pig is ahead or behind the pigs wean weight efficiency and determine how this will affect wean-to-finish expectation. A one-pound difference in wean weight, at the same age, can have a

2% impact on ADG and a 1% impact on FCR, both driving overall production efficiency. Understanding the wean weight efficiency from the sow farm also allows management to adjust starting protocols based on the pigs being raised. In addition to wean weight efficiency, it is important to monitor overall thriftiness, the health portfolio of the pigs and the consistency of the wean group. All this will lead to a better understanding of the pig being received and how to tailor management strategies to the individual pig.

3.2 Feed Budget

Working with a nutritionist will provide an adequate diet and a feed budget to go along with it. However, it is up to management to monitor and maintain that feed budget. Without the ability to meet the feed budget, overall pig efficiency will be impacted, and nutrient availability and/or feed budgeting may need to be adjusted. A challenge to the industry is to understand, in real time, if pigs are following the feed budget and if not, what the causes for this may be. The diet itself or access to the diet can be to blame if the feed budget is not being met. These items can include space (floor, feeder, water), health, quality of the wean pig, feeder management and feed deliveries.

3.3 Mortality

Mortality is embedded in many efficiency calculations and is an important consideration. As a pig grows, it continues to consume more feed, of which we do not get paid for if that pig becomes a mortality. A 1% mortality at weeks 4 – 10 weeks placed can impact FCR by 0.5%, at 11 – 14 weeks placed can impact FCR by 1%, and at 15+ weeks place can impact FCR by 1.5%. Not only do we need to reduce mortality overall to impact FCR but also understand the mortality curve during a growing period to focus on efforts around where the majority of mortality occurrence is. Unfortunately, mortality is a necessary part of the business, but if most of it can happen early in the growing period, overall efficiency will be improved.

3.4 Thermal Comfort

Pig comfort (both in terms of temperature and air quality) can impact production performance. Pigs have a thermoneutral range where they are the most efficient for lean tissues accretion and it is best if we can keep them in this range. Chilled pigs will consume more feed, but the energy will go towards maintaining body temperature rather than converting feed into muscle or fat. Similarly, hot pigs will reduce their feed intake, impacting not only ADG but also FCR while trying to stay cool. In general, a pigs thermoneutral zone is between 7.5°C and 30°C, but is highly dependent on the age and size of the pig as a smaller/younger pig has a much more narrow and warmer range than a larger/older pig. It is also important to consider strategies like mats and brooders to lower the over temperature of the barn, while still maintaining pig comfort, as this will drive energy efficiency. In terms of air quality, we should maintain humidity at <65%, CO_2 <3,000ppm, H_2S <5ppm, NH_3 <20ppm, and CO <30ppm.

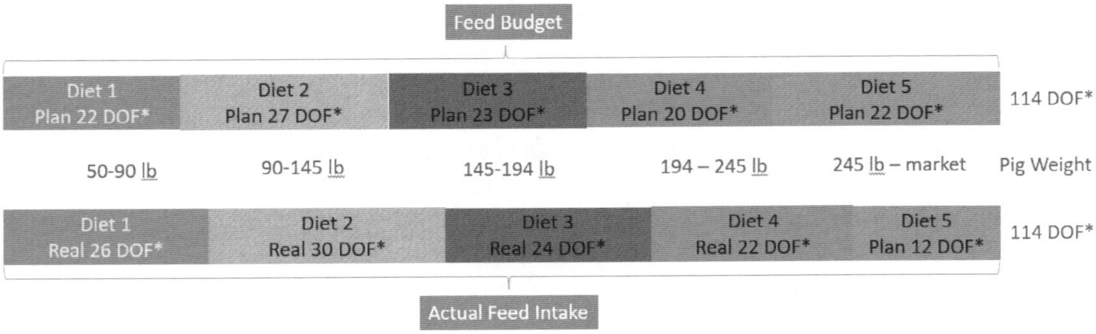

Figure 3. Feed intake and budget.

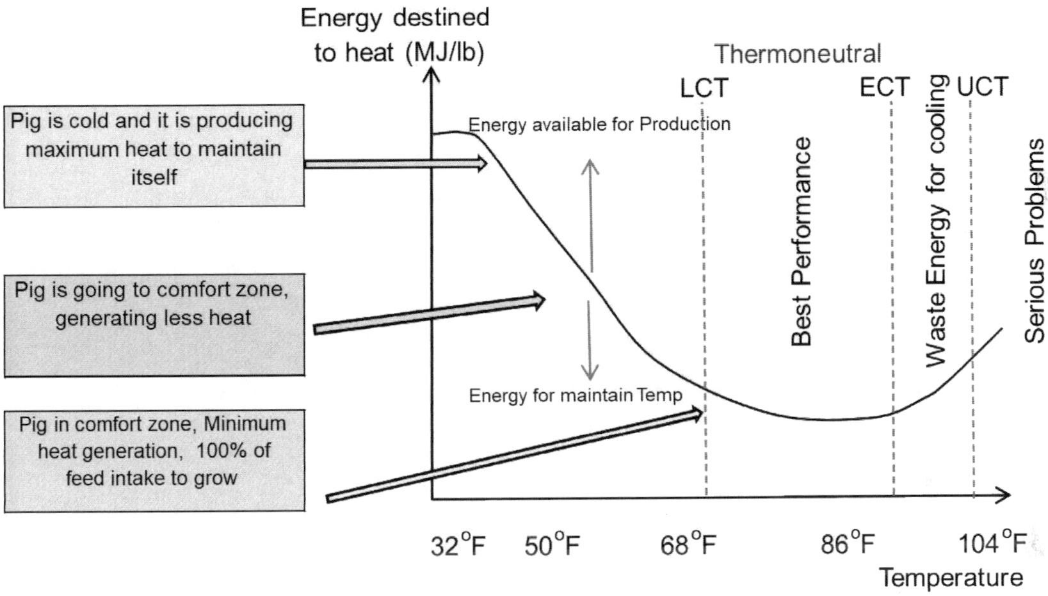

Adapted, John Gadd, 2005

Figure 4. Energy to heat.

4. Summary

To wrap up, there are many considerations when it comes to pig production efficiencies, while considering the common ones, we must also consider the over-looked areas. With each decision we make, the is a potential positive and negative result. Many factors mentioned throughout this manuscript work together and it is our job to create a balance. To drive efficiency, we first must start with good genetics to set the potential, from there we have to drive wean weight efficiency from the sow farm and then we can begin to impact efficiencies at the wean-to-finish level, including, but not limited to, diet adequacy, nutrient access and management, space, mortality, thermal comfort and health. Many factors impact production efficiency by 1% - 2% and these can likely be additive and therefore need to be heavily discussed.

References

Goncalves et al. (2016) Fact Sheet – Feed efficiency adjustments to compare close-outs in finishing pigs.
Journal of Swine Health and Production. 25.2: 73-75.

Gonyou et al. (2006) Application of broken-line analysis to assess floor space requirements of nursery
and grower-finisher pigs expressed on an allometric basis. Journal of Animal Science 84:229-235.

Lerner et. al. (2020) Effects of space allowance and marketing strategy on growth performance of pigs
raised to 165kg. Oxford University Press.

PIC Wean-to-Finish Guidelines Manual. 2019

Opportunities for Improving Feed Quality and Consistency While Reducing Feed Costs for On-Farm Feed Mills

Natalie Litvak, Jose Landero, and Malachy Young

Gowans Feed Consulting, 1811 19th Av., Wainwright, AB, T9W 1L2.
Email: malachyy@gowansfeedconsulting.ca

In today's competitive market, for pork producers to remain profitable they must work to reduce input costs. The greatest cost of production in a swine operation is feed, accounting for up to 70% of the total cost of production. Thus, focusing on feed cost, feed manufacturing and processing is crucial in ensuring profitability. There are many ways to achieve this, including ingredient selection as well as ingredient and feed processing. In Canada, many farms manufacture feed on-farm, capturing substantial cost savings when using their own or locally produced grains/pulses compared with sourcing the feed from commercial feed mills which can be a long distance from farm locations and involves extra costs for trucking. By focusing on ingredient selection as well as ingredient and feed processing on-farm, producers can reduce feed cost while optimizing animal performance, ensuring they remain profitable.

1. Advantages and Disadvantages of On-Farm Feed Milling

While commercial feed manufacturing can offer the benefit of a larger selection of ingredients, and pelleting technologies that improve the digestibility of feed ingredients, the use of on-farm feed mills offers significant potential cost savings.

For on-farm feed mills there is typically much lower movement cost associated with delivering feed to storage bins at the farm. Feed is often blown or delivered on-site, while the distance travelled from commercial mills can result in freight rates of $10-40/Tonne. Some crops produced on farm can be used when the opportunities present the best value for on-farm produced crops, thus adding further value to those ingredients. Having an extra bin on-farm for one or more opportunity ingredients can also offer substantial feed cost savings. On-farm mills typically manufacture mash feeds, which can limit inclusion of certain ingredients based on bulk density and flowability concerns. On-farm feed mills can also be limited in terms of bin space, and thus a smaller selection of ingredients can be used.

2. Ingredient Quality and Selection

Determining ingredient quality using visual appearance and analysis testing prior to formulation and feed manufacturing is important for ensuring quality feed regardless of feed manufacturing facility. A robust quality control program should be implemented working with your nutritionist. Every feed mill should have a manual outlining the ingredient purchasing specifications clearly laid out and both the feed mill manager and receiving staff should be familiar with these. Representative samples of ingredients should be collected prior to unloading to determine if the product is acceptable to unload. Visual inspection, smell, moisture, and select toxins can be run to verify the product meets specification. Once approved, further testing can be performed based on frequency of use of the ingredient, degree of problem and/or variability of the ingredients. Testing can include on-site or commercial laboratories. Accurate nutrient determination allows for more optimal diet formulation and can help reduce feed cost. It is of utmost importance that samples of all bulk ingredients being received into the mill be taken and retained for at least 3 months after use, as a fallback in the event that further testing is required at a later date. Samples should be clearly labelled with product name, supplier, and date received and stored in sealed totes at the mill. A receiving log of all bulk and bagged ingredients should be kept which indicates the ingredients received into the mill, what the truck hauled prior as well as the hauler info.

The ability to fine tune the diets in response to ingredient availability, price changes, or nutrient profile is much simpler when using micro ingredients compared to macro premixes. A micro bin system can provide savings based on the ability to custom formulate each batch, by more closely matching the nutrient requirements for each stage of production. In doing so, you avoid the over supply of expensive ingredients in some formulas and under supply in others. Typically, this system will pay itself off in 1-2 years depending on the operation.

3. Ingredient and Feed Processing

When manufacturing feed, it is important to consider feed quality as this provides opportunities to meet nutritional goals and minimize feed costs. This includes particle size of ingredients to be processed and the final feed, uniformity of the feed, and feed form.

3.1. Particle Size

Cereal grains are ground to increase nutrient digestibility by breaking the protective outer layer of the grain to allow access for digestive enzymes. The smaller the particle size, the greater the surface area for digestion. Factors affecting particle size include moisture content, ingredient to be processed (higher fibre ingredients are harder to grind), mill type and setting and mill maintenance.

For every 100-micron reduction in particle size, we see a 1.2% improvement in feed efficiency (Dritz and Hancock, 1999), thus optimizing particle size is critical for reducing feed costs. Recommended particle size for swine diets is in the 500-700 microns range and moving to this level from the more common 900-1100 microns found on many farms can produce savings of more than $3 per pig marketed. At the same time, too many fines in a ration (<500 microns) can increase the amount of dust in a facility, lead to challenges with flowability on farm and even contribute to stomach lesions and ulcers. For those mills where ingredients are pre-ground and stored, it is important to empty the pre-grind bins regularly to ensure that there is no buildup of fine material which can reduce the particle size of the final feed. With proper equipment, such a controlled grinding system, air system to pull the grain through, and variable frequency drives (VFD), optimal particle size can be achieved to maximize performance and feed efficiency. There is particle size testing equipment with 3 sieves that can be used on-farm for monitoring the grind size of ingredients and complete feed on site. This is an important tool to track the weekly particle size to determine if changes need to be made.

3.1.1. Hammer Mills vs. Roller Mills

Hammer mills reduce the particle size of ingredients by impact, causing grains to shatter. This creates a larger number of fine particles and produces a less uniform particle size distribution. Factors affecting hammer mill particle size include hammer tip speed, hammer pattern, hammer setting, screen hole diameter and whether an air assist system is used. These systems are generally lower cost to purchase and require less maintenance than a roller mill, though screens and hammers should be checked daily, magnets should be cleaned daily, and hammers should be rotated every 2-4 weeks depending on mill throughput.

Roller mills reduce particle size by crushing or shearing, producing less fines and therefore a more uniform particle size. Particle size is influenced by the number of roll pairs, the gap between rolls and the speed of the rollers. These systems consume less energy when producing particle size greater than 400 microns compared to the hammer mill, however they are more expensive upfront and have higher maintenance costs. Daily checks for gap width are required, and rolls should be re-corrugated at least annually, sometimes 3-4 times a year depending on wear. It should be noted that roller mills are not suitable for grinding fibrous materials (e.g. oats, barley).

3.1.2. Air Assist Systems

Air assist systems are commonly installed in combination with hammermills to aid with removing sized particles from the grinding chamber by creating a slight negative pressure on the outside of the screen surface. The rate at which the air assist system is operating can impact the final particle size by adjusting the time materials spend in the grinding chamber. As more air passes through the grinding chamber, sized particles will be removed from the chamber faster before further reduction in particle size occurs. Applying air flow through a hammer mill will result in more efficient operation, less regrinding, heat and moisture loss. Regular maintenance of these systems on commercial and farm feed mills is very important to ensure they are pulling the correct amount of air to achieve the grind size and uniformity being targeting. Venting of the air system is important to an external bin or storage unit to avoid a bunch of dust created inside the mill and more importantly to maintain the correct negative pressure in the grinding system.

3.1.3. Variable Frequency Drives (VFD)

Variable frequency drives (VFD) can be used to adjust the hammer tip speed. Increasing the speed results in finer grind, while decreasing the speed results in coarser grind (Saensukjaroenphon et al., 2017). This is especially important when grinding ingredients with different fibre characteristics, as the VFD can better achieve target grind size and potentially reduce the amount of very fine material produced. This also enables particle size adjustments to be made externally without requiring screens to be changed, reducing idle time in the mill.

3.2. Feed Uniformity

Uniform feed ensures that each animal receives daily all the nutrients essential for optimum performance and is achieved with proper mixing time to ensure the distribution of these nutrients. Groesbeck et al. (2007) demonstrated that as mixing time increased, and feed was more uniform, and ADG and G:F were significantly improved in nursery pigs. Uniformity of the feed is impacted by factors such as particle size and bulk density. Thomas et al. (2014) evaluated the effects of bulk density on mixing uniformity and found a high fiber diet (30% DDGS and 18% wheat midds) with a low bulk density required a mix time four times longer than a corn-soybean meal-based diet to achieve a mixer CV <10%. Thus, ensuring adequate mix time based on formulation is necessary to optimize feed efficiency and reduce feed cost. Mix time will also vary based on mixer type; horizontal vs. vertical, ribbon vs. paddle (Table 1). Horizontal mixers have been the standard of commercial feed mills, and these can be paddle mixers, ribbon mixers or a combination of both. Paddle shaped mixers tend to be the most efficient when it comes to overall mixing and are easier to maintain, as paddles can be easily replaced and repositioned. Ribbon mixers use long twisted pieces of metal that spin through the machine. The shearing action is ideal when it comes to getting rid of lumps or when adding liquids to the feed. However, this shearing can further grind down feed, reducing particle size. It also requires more maintenance and cleaning, and ribbon blades are harder to replace as they run the length of the mixer.

Other considerations include the order of ingredient addition and scale accuracy as these will affect uniformity. Ingredients should be added to the mixer from largest to smallest inclusion to ensure that micro ingredients are properly mixed. Scale accuracy ensures that ingredients are added in the correct amounts; under-addition can lead to poor performance whereas over-addition can lead to diluted nutrients and added cost. Reviewing dosing accuracy (comparing target vs. actual weight of each ingredient added in a diet) is a common approach to determine scale accuracy, however it is important that all scales in the mill are calibrated by an external company at least once per year to verify they are weighing accurately. Mixer uniformity testing should be conducted at least annually, though CFIA requirements only dictate every one to three years depending on risk profile.

3.3. Feed Form

Pelleting feeds provides benefits from improved palatability and flowability, to reduced ingredient separation and decreased feed wastage if pellet quality can be maintained. However, pelleting comes with an added cost, thus, the benefits realized from improved animal performance must outweigh this cost. Pellet quality varies with diet formulation, pelleting conditions, and feed movement. For example, complex nursery diets require lower conditioning temperatures than a finishing diet, however, this increases the frictional heat as there is less steam acting as a lubricant, which requires more energy, decreases throughput and increases die wear. Longer conditioner retention times can improve pellet quality by increasing heat and moisture absorption, however, this again decreases mill throughput and can potentially reduce some nutritional value in the feed. It is important to measure both pellet durability (using the pellet durability index, PDI) at the mill as well as the pellet fines at the feeder level. A key consideration with pelleting is the capital cost per tonne produced annually. On-farm pelleting is rarely justifiable due to the high capital cost and on-going energy and labour costs associated with running the system.

References

Dritz SS, Hancock JD. Grain Particle Size: Influence on Swine and Poultry Performance and Practical Methods for Monitoring. American Soybean Association Technical Bulletin. 1999. Pp 632

Groesbeck CN, Goodband RD, Tokach MD, Dritz SS, Nelssen JL, DeRouchey JM. Diet mixing time affects nursery pig performance. J Anim Sci. 2007 Jul;85(7):1793-8. doi: 10.2527/jas.2007-0019. Epub 2007 Apr 12. PMID: 17431042.

Saensukjaroenphon M, Evans CE, Sheldon KH, Jones CK, Paulk, CB, Stark CR. The Effect of Hammermill Screen Hole Diameter and Hammer Tip Speed on Particle Size and Flowability of Ground Corn, Kansas Agricultural Experiment Station Research Reports (2017) Vol. 3: Iss. 7. https://doi.org/ 10.4148/2378-5977.7505

Thomas LL, Stark CR, Fahrenholtz CH, Coble KF, Hayashi S, Saensukjaroenphon M, Woodworth JC, and Goodband, RD. Effects of diet bulk density on mixing uniformity, Kansas Agricultural Experiment Station Research Reports 2014: Vol. 0: Iss. 10. https://doi.org/10.4148/ 2378-5977.6887

BACK TO BASICS
START PIGS RIGHT
Getting pigs started right is more important than ever with today's challenges. With a proven track record, AP 920 plasma can help you get the job done quickly, easily and effectively. Sometimes getting back to the basics is the best solution.
100% Bovine | 100% Porcine | UV Processed
APCproteins.com
WATCH THEM thrive
APC

The MAXIMUS Solution
A smart integrated system that combines productivity and animal welfare.
MAXIMUS
An Ingersoll Rand Business
Equipment automation according to customized parameters:
- Temperature
- Humidity
- CO2
- NH3
- Ventilation
- Feeding
- Water
- Etc.
Instantaneous alarms that flag any irregularities
Real-time data import and customizable report generation
Herd management - Trend analysis at the individual and group levels
Synchronization of production tracking for all buildings and sites
maximus-solution.com
1 877 445-6556
Learn more

Sow Nutrition & Management With its Impact on Sow Mortality

Laura L Greiner

Iowa State University, 109H Kildee Hall, 806 Stange Rd, Ames, IA 50011. greinerl@iastate.edu

1. Introduction

Sow culling rate in the United States is currently around 47% with sow mortality increasing from 7 to 15% in the last seven years (Supakorn et al., 2018). A majority of sow mortality occurs within the two weeks pre and post-farrow. Both sow mortality and longevity are tied together and are influenced well before the gilt is first bred. One of the primary reasons for sow removal in a herd is due to lameness. A recent evaluation conducted by Chipman et al. (2018) demonstrated that 29% of sow mortality is associated with lameness and 21% are associated with pelvic organ prolapse. The purpose of this paper is to provide some insight into potential nutritional and management strategies to reduce sow mortality.

2. Gilt Development

The process of growth results in small amounts of inflammation within the joints. Over time, this inflammation and any other joint injury could result in future arthritis or other types of lameness. Bone maturation in a commercial gilt today is not complete until after she has had her first farrowing. In addition to adequate mineral supplementation to support bone maturation and ligament strength, consideration should be made to the joint. Becker and Greiner (2022) determined that feeding lower energy diets during the developmental stage of gilt growth, certain proteins associated with inflammation were reduced. Finally, body condition of the gilt is critical. A high percent of gilt developer units will allow gilts to have access to full feed until the time of the first heat expression. Mammary development will occur from 90 days of age through puberty, late gestation and lactation (Farmer and Hurley, 2015). During the grow-finish phase of gilt development, most of the fat deposition will occur subcutaneously. However, over-conditioning will result in fat deposition to occur within the mammary gland, which can impair milk production (Hilson et al., 2004; Rasmussen, 2007).

3. Gestation

The use of open pen housing has been slowly increasing over the last 10 years. While group housing has advantages, the hierarchy establishment period can create opportunity for musculoskeletal injuries. It is important for producers to manage the mixing period to minimize these events as they could have long-term consequences (Knox et al., 2014). Body condition and proper management to prevent prolapses must also have close observation and action within this timeframe. Ross et al. (2022) discussed in recent research that females with higher perineal scores in late gestation had an increased risk of prolapse.

4. Lactation

Proper vitamin and mineral supplementation throughout the sow's reproductive cycle is important. Overtime, the tissues levels in sows decline (Mahan and Newton, 1995). However, due to advances in laboratory analysis and genetic changes within the sow herself, the understanding of the sow's vitamin and mineral requirements may require updating. In a recent study by Greiner et al. (2022), some vitamin and mineral concentration ranges differ from previously published values (Puls 1994a, 1994b). Greiner et al. (2022) determined that serum vitamin A (0.03-0.32 ppm) are lower than previously published values (Puls,1994a). In addition, serum selenium levels (0.133-0.355 ppm) were also reduced relative to previously published literature (Puls, 1994b). These findings lead to the question of why are these levels lower and could they be contributing to a shorter lifetime productivity.

While much work has been done over the years to determine the lysine and other amino acid requirements (Greiner, 2019), work is still needed to better assess the branch chain amino acids and how they may interact with each other both during absorption and their relationship to sow feed intake (Holen et al., 2022). Furthermore, additives that can improve the sow's ability to manage heat stress need to continually be evaluated as there are seasonal increases in sow mortality associated with summer heat (Greiner et al., 2014).

Researchers today are also working towards better understanding the microbiota of the sow and how that not only could impact offspring performance but how this may change the sow's performance as well. In a recent study conducted by Nair, et al. (2019), it was found that when the sow's microbiota contains high levels of Ruminococcus_2, there is a reduced incidence of lung lesions when the animals are exposed to *Mycoplasma hyopnuemoniae*. Additional research needs to be conducted to determine how to best nutritionally increase certain organisms that could have protective benefits to the animal during health challenges.

5. Conclusion

Nutrition can have a significant impact on sow mortality. Nutrition programs to maximize sow longevity start in the gilt developer unit and continue throughout a sow's lifetime. While some of the nutrition programs focus on diet additives or adjustments, other aspects of improving sow longevity come from managing the nutrition programs such as optimizing body condition.

References

Becker, S. and L. Greiner. (2022) An Investigation into the Role of Dietary Essential Fatty Acids Ratios and Energy Level on Growth Performance, Inflammation, and Joint Health of Grow-finish Pigs. *Journal of Animal Science*, Volume 99(Supp1):63–64.

Chipman, A., Z. Kiefer, C. Rademacher, C. Johnson, K. Stalder, A. Johnson, A. Keating, J. Patience, N. Gabler, D. Linhares, K. Schwartz, S. Millman, S. Schmitz-Esser, Gustavo Silva, J. W. Ross. (2018). An industry-wide approach to better understand the putative causes of pelvic organ prolapse in sows. https://www.youtube.com/watch?v=mLVEPUaOMvY

Farmer, C. and W. Hurley. (2015). The gestating and lactating sow. **ISBN:** 978-90-8686-253-5

Greiner, L. (2019) "Mammary uptake of metabolites and requirements of energy and amino acids of hyperprolific sows" in the Novus book Nutrition of Hyperprolific Sows.

Greiner, L. L., S. Elefson, S. Radke, C. Hagen, D. Humphrey, and S. Becker. (2022) A survey of vitamin and trace mineral ranges for diagnostic lab reporting from conventionally raised swine. *J Swine Health Prod.* In press.

Greiner, L., J. Remus, D. D. Hall, R. D. Boyd, R. J. Harrell and J. Connor. (2014). Abstract 146. The impact of natural betaine to lactating sows during periods of heat stress. Midwest American Society of Animal Science. March 2014.

Hilson, J.A., K. Rasmussen, C. Kjolhede. (2004) High prepregnant body mass index is associated with poor lactation outcomes among white, rural women independent of psychosocial and demographic correlates. *J Hum Lact.* 20(1):18-29.

Holen, J., M. Tokach, J. Woodworth, J. DeRouchey, J. Gebhardt, E. Titgemeyer, and R. Goodband. (2022) A meta-regression analysis to evaluate the influence of branched-chain amino acids in lactation diets on sow and litter growth performance, *J Anim Sci,* 100 (4): skac114.

Knox, R., J. Salak-Johnson, M. Hopgood, L. Greiner, and J. Connor. (2014) Effect of day of mixing gestating sows on measures of reproductive performance and animal welfare. *J Anim Sci.* 92:1698-1707.

Mahan D.C. and E. A. Newton. (1995) Effect of initial breeding weight on macro- and micromineral composition over a three-parity period using a high-producing sow genotype. *J Anim Sci.* 73(1):151-8.

Nair, M., T. Eucker, B. Martinson, A. Neubauer, J. Victoria, B. Nicholson, and M. Pieters. (2019) Influence of pig gut microbiota on Mycoplasma hyopneumoniae susceptibility. Vet Res. 28;50(1):86.

Puls, R. *Mineral Levels in Animal Health: Diagnostic Data / by Robert Puls.* 2nd ed. Clearbrook, B.C., Canada: Sherpa International, 1994. Print.

Puls, R. *Vitamin Levels in Animal Health / by Robert Puls.* 2nd ed. Clearbrook, B.C., Canada: Sherpa International, 1994. Print.

Ross, J., Z. Kiefer, and J. Studer. (2022). Pelvic organ prolapse: On-farm considerations. National Hog Farmer. Pelvic organ prolapse: On-farm considerations (nationalhogfarmer.com)

Rasmussen, K. M. (2007) Association of maternal obesity before conception with poor lactation performance. *Annu Rev Nutr.* 27:103-21.

Supakorn, C., J. Stock, E. Garay, A. Johnson and K. Stalder. (2018). Lameness: a principle problem to sow longevity in breeding hereds. *CAB Reviews* 13:23.

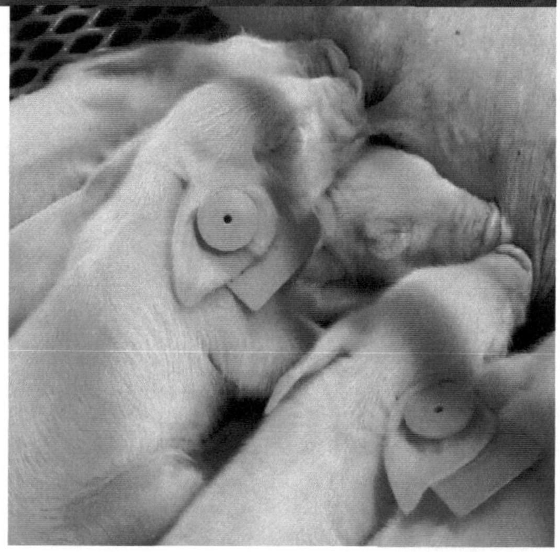

Fostera Gold PCV
Fostera Gold PCV MH
THREATS ADAPT
SO SHOULD YOUR PROTECTION
BROADEST ANTIGENIC AND
LONGEST LASTING PCV2 PROTECTION
zoetis
All registered trademarks are the property of Zoetis or its licensors, used under license by Zoetis Canada Inc. ©2021 Zoetis Canada. All rights reserved.

Swine Heaven in 2050 - Building the Future of Pig Housing

Sébastien Fournel

Department of Soils and Agri-Food Engineering, 2425, Agriculture Street, Université Laval, Quebec City (QC), Canada, G1V 0A6
Email: sebastien.fournel@fsaa.ulaval.ca

1. Introduction

Over the last twenty years, the Canadian swine industry has undergone major changes, evolving from small family businesses to intensive farming. Process standardization and automation led to an improved productivity, allowing meeting an increasing worldwide demand. However, this industry change has caused higher greenhouse gas (GHG) emissions due to increased animal density and expansion of cultivated lands. Moreover, the consumer is increasingly aware of the environmental impact of modern agriculture and now advocates for enhanced animal welfare. Consequently, the hog industry not only needs to maintain a high productivity but must also meet rising societal expectations. In response to those numerous challenges, many researchers around the world aim at providing better livestock conditions reuniting health, welfare, protection of the environment and profitability. The goal of the conference is to present various concepts that could be implemented by 2050 to address sustainability issues.

2. Towards More Natural and Sustainable Pig Farming

In response to societal demands in terms of ethical practice, the pig industry should provide conditions closer to those found in the nature for animals. Because pigs are gregarious, curious and very intelligent animals, particular attention should be paid to natural behaviors such as socialization, resting, exploration, waste elimination and thermoregulation. When it is possible, pigs prefer to explore, play, rest and defecate outside during daytime, and sleep inside the building at night (Stolba and Wood-Gush, 1989). In addition, providing an adequate density and enrichments, such as straw, significantly reduces aggressive behaviors and injuries (Studnitz and al., 2007). A layout example integrating such key points is shown in Figure 1. The presentation will develop the idea of a swine heaven along six themes: (1) closer to nature; (2) comfortable surfaces; (3) healthy environment; (4) circular and synergistic production cycles; (5) carboneutrality; and (6) smart farms.

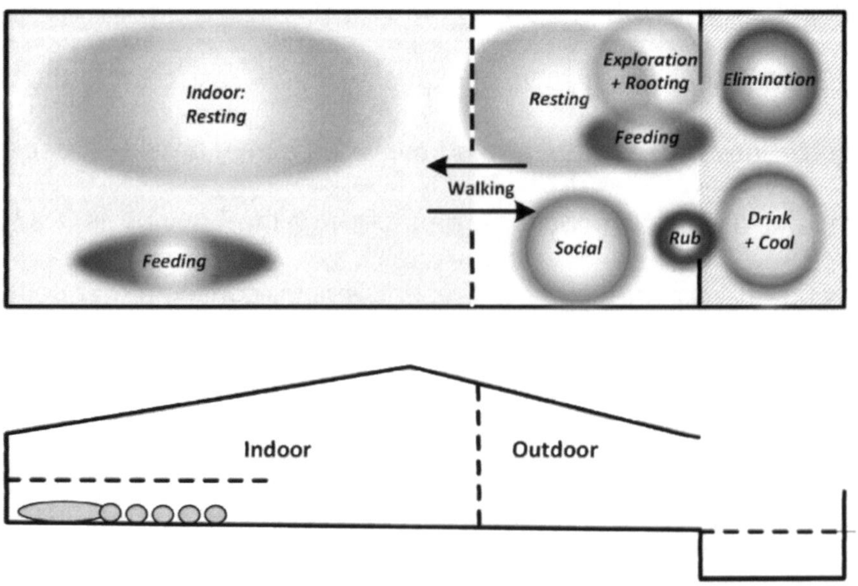

Figure 1. An example of a layout with outdoor access and different sections for various activities. (Wimmler and al., 2022).

2.1 Closer to Nature

In modern constructions, sows tend to be kept in groups during a part of the gestation and eventually during lactation. Less contention allows the expression of maternal behaviors, without affecting the performance (conception rate, litter size, pigs born alive per litter and stillborn per litter) or longevity (Anil and al., 2006; Ramonet and al., 2019; Vande Pol and al., 2018). In addition, different feeding systems and interior layouts can limit food competition. Growers can also benefit from larger areas with outdoor access and enrichment: studies showed more exploration behaviors, less aggression, improvement of animal welfare and higher slaughter weight (Liorancas and al., 2006; Nannoni and al., 2019). Moreover, researchers are currently assessing the potential benefits of greenhouse barns, providing natural light, and creating shade using vegetation (Leso and al., 2017).

2.2 Comfortable Surfaces

Slatted floors increase the risk of injuries and lameness in pigs (KilBride and al., 2009; Ludwiczak and al., 2021). The ideal flooring should be soft, comfortable, durable and quickly filter urine and retain feces. Artificial floors with gutters, which can drain urine out of the building, and synthetic mats, which can be cleaned automatically by using a robot multiple times a day, can also reduce ammonia (NH_3) emissions (Wageningen UR Livestock Research, 2011, ID Agro 2022).

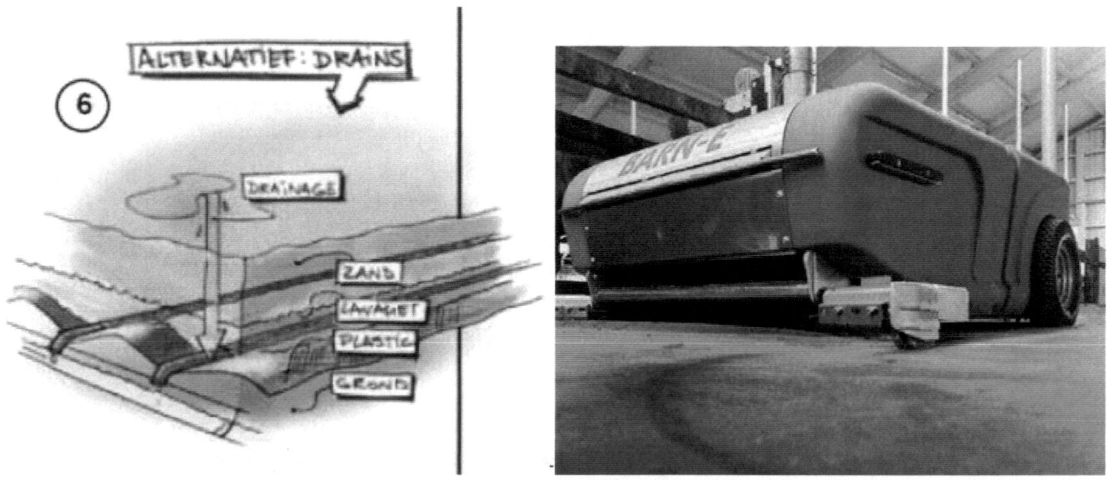

Figure 2. Alternative synthetic floor with four filtering layers and manure cleaning robot (Wageningen UR Livestock Research 2011; ID Agro 2022).

2.3 Healthy Environment

The best ventilation system seems to be a central exhaust system. Its advantage is that the air can be treated at a central point before being released into the environment. Therefore, it can reduce up to 92% of dust, 75% of odors, and 77% of NH_3 emissions. This improves air quality in the building as well as at the exhaust to preserve animal and worker health and improve cohabitation (Belzile and al., 2014). In addition, the air is easier to heat or cool before it reaches the rearing rooms to ensure appropriated ambient conditions.

Moreover, in a climate change context, cooling methods will become more and more important. This can be achieved with evaporation panels on walls, better attic insulation, adequate adjustment of the air inlet, recirculation fans, misting and sprinkling of the animals.

2.4 Circular and Synergistic Production Cycles

It is crucial to identify sustainable livestock and manure management practices, including manure treatment technologies. In the building, frequent scraping and under-slats separation of feces and urine helps to conserve nitrogen for a better agronomical value (Varma and al., 2021; Godbout and al., 2010). It is also possible to mechanically separate manure into a liquid fraction rich in nitrogen and a solid fraction that can be composted or transformed into biofuel (pyrolysis or biomethanization).

During storage, a covered pit can help to limit the dilution of manure through precipitations and also limits the volatilization of nitrogen. Furthermore, better storage practices reduce the costs related to the management of fertilizers and manure and, in certain scenarios, improved grain sales due to a better fertilizing value of the manure (Fournel and al., 2019).

Insects and/or fish could be produced in multi-trophic production systems with poultry, pigs and/or cattle. Their excrement could be used to produce insect larvae or earthworms that serve as food for terrestrial and aquatic livestock, reducing the need for fishmeal or feed produced with fertilizers and fossil fuels. Also, the residues from invertebrate production can be carefully returned to cultivated fields and pastures (Kronberg and al., 2021).

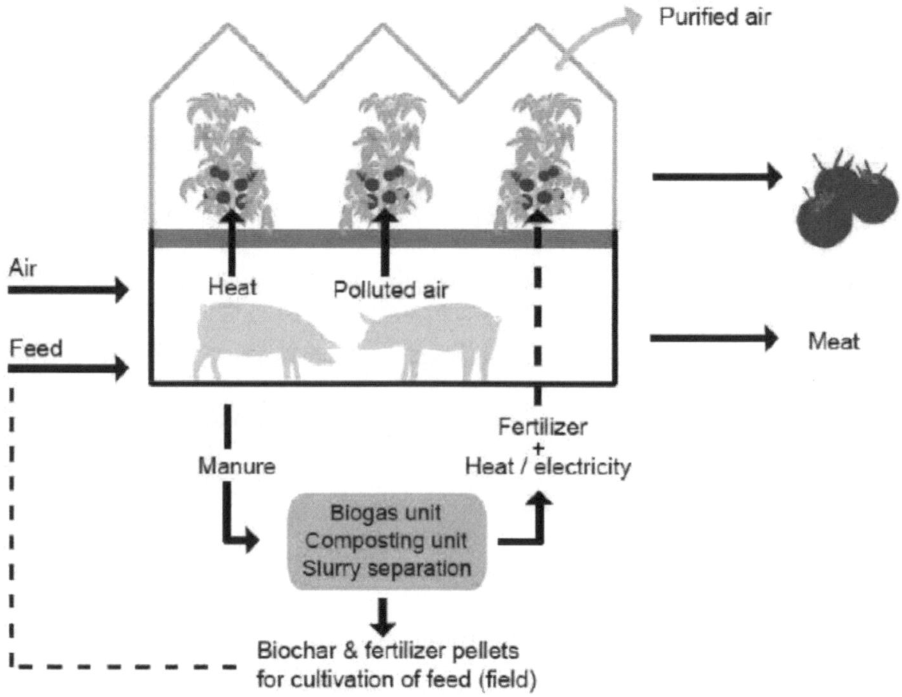

Figure 3. Pig City: An example of circular and sustainable farming in Denmark (Gottlieb Paludan, 2014)

2.5 Carboneutrality

The most effective way to reduce farm building emissions is via manure separation. This can be achieved through a filtering synthetic floor, an underslat V-shaped scraper or a mechanical separator as previously mentioned. The solid fraction can be dried, composted or converted into methane.

Fossil fuel consumption can be reduced by implementing renewable energy sources such as solar panels, solar wall or geothermal energy. Heat exchangers can also reduce propane usage, resulting in annual savings (IRDA, 2008).

2.6 Smart farms

Technological progress makes precision farming more advanced and available. Smart sensors, cameras and microphones acquire a huge amount of data that can inform the producer in real time of the health and well-being of their animals. Mathematical models make it possible to automatically adjust the environmental conditions and detect abnormalities, thus optimizing production.

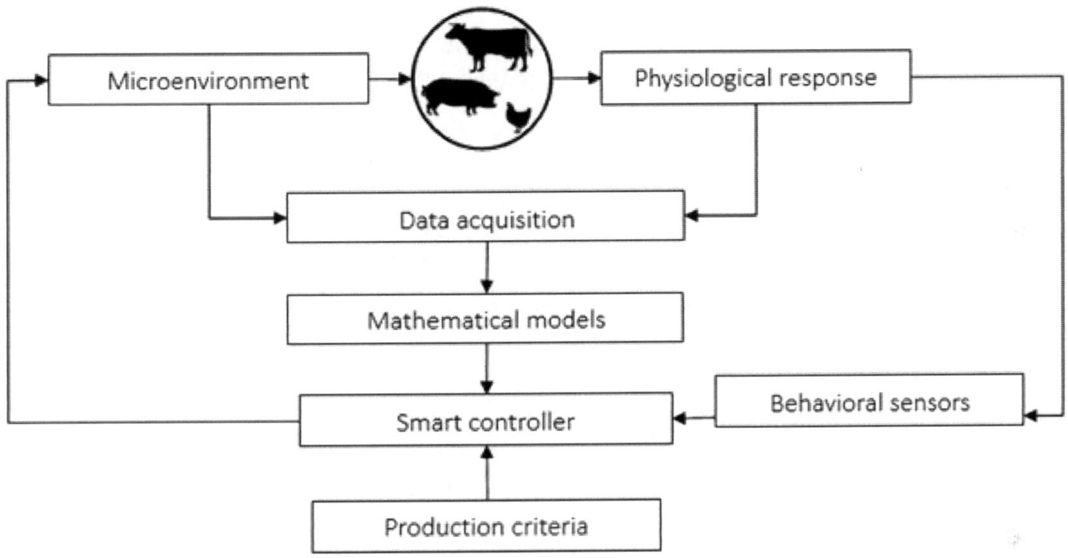

Figure 4. Components of precision livestock farming (Fournel and al. 2017)

3. Conclusion

In summary, the future of pig housing lies with an optimized production cycle to meet the rising global demand while reducing its environmental footprint and improving animal welfare. To achieve this, hogs should be raised in more natural and comfortable conditions where a central exhaust system could condition and purify the air. The implementation of solid-liquid separation would allow an optimal use of pig manure fractions that can be used in a circular system to produce renewable energy. Finally, precision livestock farming principles could be applied to improve productivity and animal welfare. This sustainable concept may seem idealistic, but "pig cities" already showed promising results in Europe and could be adapted to the Canadian context.

References

Belzile, M. and al. (2014). Innovative air treatment unit for swine exhaust air. *Annual International Meeting 2014*. **7**. 4874-4882.

Fournel and al. (2017). *Biosystems Engineering*, **155**: 96-123.

Fournel and al. (2019). *Transactions of the ASABE*. **62**(4): 973-984.

Gottlieb Paludan. (2014) Pig City flyer. https://www.gottliebpaludan.com/en/project/future-farming-pig-city

ID Agro. *High Welfare Floor*. https://www.idagro.nl/producten/high-welfare-floor.

IRDA. (2008). *Échangeur d'air récupérateur de chaleur Polymair* [MD]. https://www.irda.qc.ca/fr/realisations/transfert-technologique/polymair/.

Kilbride, A. and al. (2009). Impact of flooring on the health and welfare of pigs, *Practice*, ***31***(8), 390–395.

Kronberg, S. L. and al. (2021). Review: Closing nutrient cycles for animal production – Current and future agroecological and socio-economic issues, *Animal*, ***15***.

Leena Anil, S. and al. (2006). Effect of group size and structure on the welfare and performance of pregnant sows in pens with electronic sow feeders. *Can J Vet Res.*, ***70***(2), 128–136.

Leso, L. and al. (2017). Evaluating thermal performance of experimental building solutions designed for livestock housing: The effect of greenery systems. *Agronomy Research*, ***15***(1).

Liorančas, V. and al. (2006). Influence of rearing space on the behavior, performance, carcass and meat quality of pigs. *Med. Weter.*, **62**(3):274-277.

Ludwiczak, A. and al. (2021). How housing conditions determine the welfare of pigs. *Animals*, ***11***(12).

Nannoni E. and al. (2019) Effects of increased space allowance on animal welfare, meat and ham quality of heavy pigs slaughtered at 160Kg. *PLoS ONE* **14**(2).

Ramonet, Y. and al. (2018). *Les truies libres en maternité : modalités de logement et de conduite, performances, conditions de travail et bien-être*.

Stolba, A., Wood-Gush, D.G. (1989). The behaviour of pigs in a semi-natural environment. *Anim Prod Sci*, **48**, 419-425.

Studnitz, M. and al. (2007). Why do pigs root and in what will they root? A review on the exploratory behaviour of pigs in relation to environmental enrichment. *Appl. Anim. Behav. Sci.*, **107**(3-4), 183–197.

Vande, K. D. and al. (2018). 19 Effect of Gestation Housing System (Individual vs. Group) on the Reproductive Performance of Sows over 6 Parities Under Commercial Conditions. *J. Anim. Sci* ***96***(S2).

Varma, V. S. and al. (2021). Dairy and swine manure management – Challenges and perspectives for sustainable treatment technology. *Sci. Total Environ.*, ***778***.

Wageningen UR Livestock Research. (2011). Prospects for bedded pack barns for dairy cattle.

Wimmler, C. and al. (2022). Review: Concrete outdoor runs for organic growing-finishing pigs – a legislative, ethological and environmental perspective. *Animal*, ***16***(1).

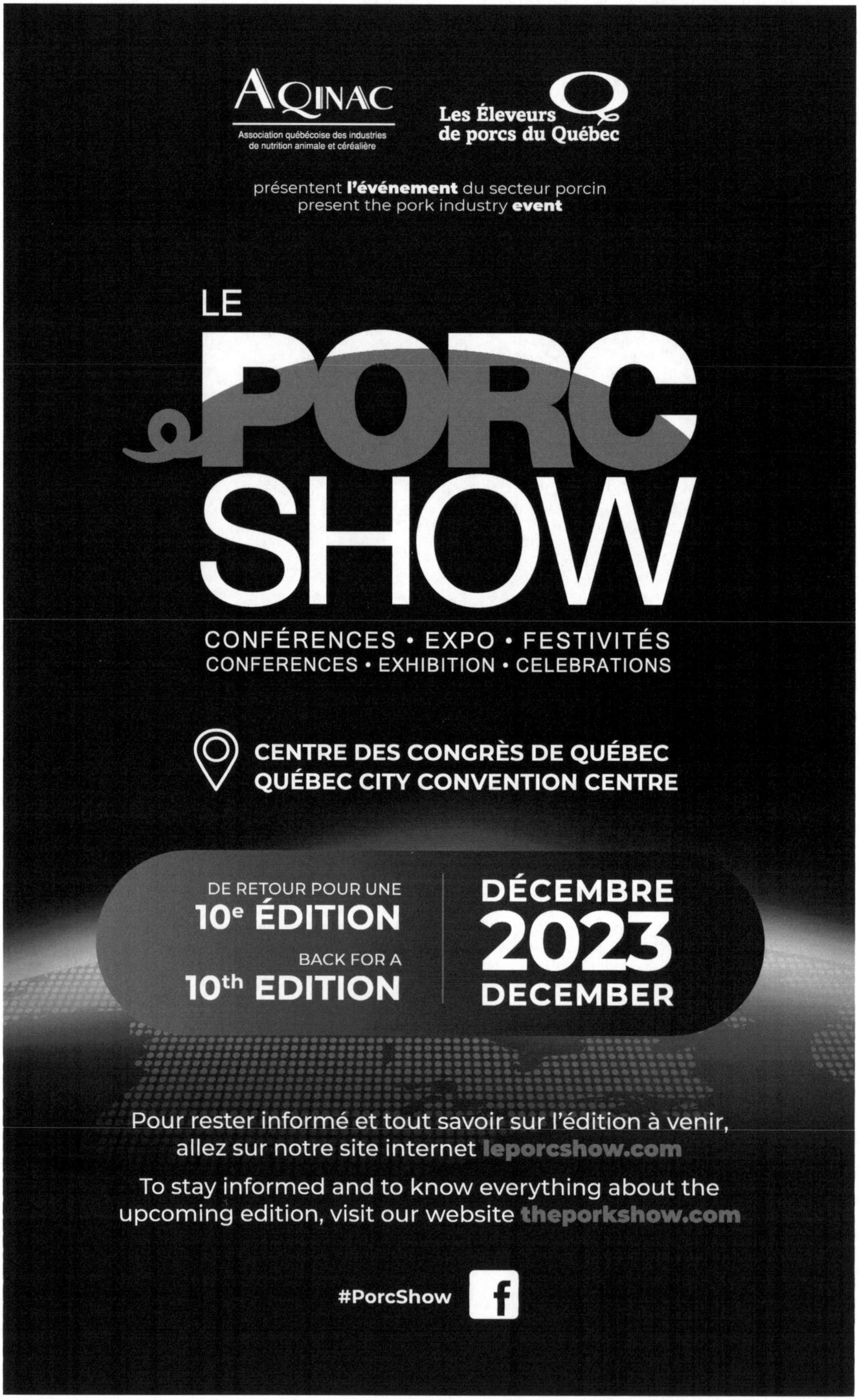
AQINAC
Association québécoise des industries de nutrition animale et céréalière
Les Éleveurs de porcs du Québec
présentent **l'événement** du secteur porcin
present the pork industry **event**
LE PORC SHOW
CONFÉRENCES • EXPO • FESTIVITÉS
CONFERENCES • EXHIBITION • CELEBRATIONS
CENTRE DES CONGRÈS DE QUÉBEC
QUÉBEC CITY CONVENTION CENTRE
DE RETOUR POUR UNE
10e ÉDITION
BACK FOR A
10th EDITION
DÉCEMBRE 2023 DECEMBER
Pour rester informé et tout savoir sur l'édition à venir, allez sur notre site internet leporcshow.com
To stay informed and to know everything about the upcoming edition, visit our website theporkshow.com
#PorcShow

Get more from your feed by
elevating
gut health
Strive for nutrient utilization
with ACTIGEN®
FORMULATED TO:
Aid in nutrient utilization
Build natural defenses
Support post-weaning performance
Optimize sow condition
To learn more,
visit alltech.com/actigen
Alltech®

Manure Management & Treatment - State of Knowledge

Stéphane Godbout, Patrick Brassard and Joahnn H. Palacios

Research and Development Institute for the Agri-environment (IRDA), 2700 Einstein Street, Quebec City (QC) Canada, G1P 3W8.
stephane.godbout@irda.qc.ca

1. Context

1.1. Introduction

Manure management is an issue in many countries having a large swine production. Due to the environmental regulation, agronomic soil characteristics, social and economic situation and climate, the challenges related to manure management are solved or studied on different ways. Over the years and the world, a lot of research projects are carried out. The scientific publication number could be an indication of the evolution of that research field over the year in the world. The peak has been reached in 2019 in Canada. In the Province of Quebec, a lot of research projects have been done at the beginning of the 2000 decade. At that time, the works were mainly driven by the phosphorus (P) regulation and odour emissions, with a light concern about ammonia emissions. Due to additional cost related to the treatment or the new practice, very few new technologies or practices have been adopted.

1.2. Social and Environmental Challenges

In Canada, manure in conventional pig farming is managed in a liquid form (Turcotte, 2022). The liquid swine manure is a good and natural fertilizer. However, the agronomic characteristics cannot be really controlled and consequently, it is difficult to optimize and synchronize the quality of the manure with the crop needs in terms of phosphorus and nitrogen needs. A better efficiency could bring to a reduction of synthetic fertilizer use.

In 2020, the Canadian agriculture sector has emitted 55 Mt of CO_2e, representing 8.2% of the total greenhouse gas (GHG) emissions from the country. A large part (40%) of these emissions come from the agricultural land. The swine manure management is responsible of around 1.72 Mt of CO2e from the building to the land.

The Canadian agriculture is the third GHG emitting sector following stationary combustion sources and transport sectors, but is the largest contributor to ammonia emissions since 93% come from the agriculture. Manure management (all species) contributes to 69% of the agricultural NH_3 emissions. For now, there is no real regulation in Canada to reduce these emissions. It is not the case in many European countries having specific regulation to reduce or eliminate these emissions.

In many countries, odour from pig production, mainly from manure spreading operation, is an important issue. Many countries, provinces or stated manage these emissions using a setback distance approach (Godbout et al., 2016). Spreading technic used, landscape or natural windbreaker and the type of manure (raw or treated) can have an impact on setback distance.

In the past, there were some concerns about the impact of liquid manure spreading on the underground and surface water. More recently, some research studies pointed out manure management as a source of bioaerosols and a source of antibiotic resistance bacteria. These emissions might have an impact on human and animal health. All these challenges highlighted the importance to improve liquid manure management and practices to reduce the impact on the environment and on the society.

2. Manure Management Practices

2.1. Separation

2.1.1.Separation as a part of management chain

This approach produces two fractions, solid and liquid. The solid has on average 30% of dry matter, rich in phosphorus and should be post treated to be stabilized or to produce other by-products. The liquid is an interesting fertilizer with a high nitrogen content. If the solid is well managed, this approach allows a direct GHG emission reduction at the storage and might reduce indirectly the emissions by optimizing the use of manure nutrients by the plants. Equipments are needed to manage both fractions, which is the main disadvantage as it increases the production cost. Separation is often the first step of many treatment systems.

Manure management in two fractions gives more opportunities of valorization as fertilizers or other subproducts.

Once separated, the liquid fraction can be used as a fertilizer or treated on farm, using technologies like ultra-filtration or reverse osmosis for a rejection in the environment. The solid fraction can be transported elsewhere for fertilisation, or to a plant for further transformation (fertilizer pellets, thermochemical process, etc.). For example, biochar production through pyrolysis or hydrothermal carbonization and its amendment to soil is a solution for carbon sequestration in agricultural land. In the decade 2000-2010, some research studies explored the possibility to use ashes from dry solid fraction (after thermochemical process) as part of the diet (van Kempen, 2003 and van Kempen et al. 2003).

2.1.2. Mechanical separation

Mechanical separators are the most used since many years over the word in animal production. With or without additives, screw, centrifugal, sieve or rollers are the main techniques used (Fournel et al., 2019). The solid fractions produced from these techniques are similar, but the efficiency, principle and maintenance cost are quite different.

2.1.3. In-barn separation

At the EuroTier Fair last November, six companies presented technologies related to the under-slat separation system (Hill, 2022). This system has the advantage to collect manure in two fractions and allows an important reduction of ammonia emissions (regulated in France). For example, Cooperl in Britanny (France) is implanting a concept with their members including an under-slat separation system "ValuTRAC". This system is already implemented on 134 farms (Turcotte, 2022).

2.2. Treatment

2.2.1. Aerobic treatment

This technique consists of injecting air in the raw manure or liquid fraction using different rates and on/off sequences. An air injected manure favorizes the solid settlement. If parameters are well controlled, this technique can reduce ammonia and methane emissions, kills pathogen, and reduces odor. Usually, the pH increases during the aeration. Very popular in Europe due the nitrogen regulation, this technique has been also investigated in Canada between 1995 and 2005. Due to the cost and no real regulation, very few aerobic treatments have been built and promotors went to bankrupt. Usually built in the storage tank, the pressure and energy needed impact the operation cost. To address theses challenges, a new technology recently patented will be in available soon (Roy et al. 2021). For a liquid fraction (from separation), the aerobic treatment is more efficient and less costly due to a lower organic matter than the raw manure.

2.2.2. Anaerobic treatment

The goal of this process is to produce biogas from digestion of manure, which is continuously mixed in anaerobic conditions. This technique assumes that raw manure is used and sometimes is enriched to produce high quality and quantity of biogas. The digestate can be used as fertilizer directly or follow the separation path. This technique is well known for the significant reduction of GHG emissions. The gas can be used for heating or to be insert in the natural gas network after cleaning and upgrading.

This system is very popular in Germany, Denmark and Sweden to produce energy (electricity or heating gas).

2.2.3. Other treatments

Some additives are available, and they usually acidify the manure to reduce GHG emissions. The additives have a variable effect and doesn't improve a lot the fertilizing properties of the manure.

At the begging of the 2000, there were many companies offering complete packages to treat the manure. Nowadays, few remain in Canada.
Solugen technology consists in a distillation of the liquid fraction after raw manure separation. The distilled water can be use on farm. The subproducts are a high nitrogen solution, rich phosphorus biosolid and potassium concentred liquid (Solugenglobal.com).

2.3. Storage

There was no real innovation regarding the storage tanks for the last 20 years. However, more data on gas emissions are available. These emissions are mainly influenced by the manure temperature and surface area (more than the volume). For example, emptying completely the tank at spring and fall significantly reduces gas emissions because the inoculum is eliminated. This practice should be implemented at large scale due to his low cost and efficiency.

Roofs (floating, rigid structure or inflated) and still used and available on the market. Certain airtight cover can be used as anaerobic digestor (low efficiency) and to produce biogas. The main benefits of these structures are: 1) reduction of water volume bringing less transportation, 2) gas emission reductions (CH_4 and NH_3), and 3) reduction of odour (avoiding complaints from the neighborhood).

2.4. Spreading

Pig slurry is in most of the cases spread using splash plate at 1 m over the soil surface, which is an important source of emissions like dusts, bioaerosols (pathogens), gas and odour. In some Canadian provinces, the manure applied on agricultural land have been targeted as responsible of the water quality degradation and lake eutrophication. All these negative impacts will bring to important changes in the way to apply manure. Unfortunately, in Canada, there was very few public research on that field, leaving the producers without real solutions. The results from previous research studies tend to recommend manure injection or incorporation for odour, gas, and runoff control (Chen, 2022). In any case, spreading should include incorporation to avoid emission and runoff combined with precision agriculture principles. Moreover, Fertilizer Canada has implemented the 4R Nutrient Stewardship (Right Source at Right Rate, Right Time, Right Place ®), which contributes to minimize the environmental impact of manure spreading.

3. Conclusion

There have been very few changes over the last 20 years, because there is not real constrain regulation and due to the additional cost related to the implantation and operation of new techniques (Turcotte, 2022). Due to public concern, the pandemic, the GHG emissions, since few years a lot of research projects are carried out on agronomic, animal welfare, sanitary status, water and air pollution and antibiotic resistance genes. These studies will bring changes in livestock industry. Few trends can be pointed out. First, manure management in two fractions (after separation at or in-barn) following by the treatment of both fractions seems to become popular, mainly in France. Then, spreading of raw manure on land will be discouraged. In all cases, raw or fractioned manure should probably be completely or partly treated in the future. Finally, for many reasons, spreading operation should include incorporation, during the operation or immediately after.

4. Acknowledgements

Thanks to the Banff Pork Seminar organising committee for the invitation. I would like to also thank Beatrice Dupont-Fortin for her help to regroup recent information's and IRDA organisation to allow me to present at this seminar.

References

Chen, Y. 2022. Private communications.

Fournel, S., S. Godbout, P. Ruel, A. Fortin, M. Généreux, C. Côté, C. Landry and D. Pellerin. Production of recycled manure solids for bedding in Canadian dairy farms: I. Solid-liquid separation. J. Dairy Science, 102:1832-1846.

Godbout, S., J.H. Palacios, S. Sakka, F. Pelletier, S. Fournel and F.X. Philippe. 2016. Inventaire et comparaison à l'échelle mondiale des approches par distances séparatrices pour atténuer les nuisances olfactives en production porcine. JRP, Paris, France, 2 p.

Hill, J. 2022. Private communication.

Roy, L., Godbout, S. and P. Brassard. 2021. Process for treating livestock waste to produce fertilizers. International patent application WO 2022/036460 A1. Published on 24 February 2021.

Turcotte, S. 2022. Private communication

van Kempen, T., B. Kaspers, P. Burnette, M. van Kempen, J.B. Koger. 2003. *Swine housing with a belt for separating urine and feces; key to flexibility?* In 2nd International Swine Housing Conference, 159-165. October 12-15. Durham, North Carolina. ASAE, St. Joseph, Michigan, USA.

van Kempen, T. 2003. *Re-Cycle: a profitable swine production system for zero waste*. Advances in Pork Production. Volume 14: 195.

New Standard
GROUP
TECHNOLOGY • TALENT • TRUST
We make group housing a success!
"Good condition of the sow means 0.5 more piglets per litter on average"
* Leading Nedap Technology
* Most Experienced
* New designs and Conversions
* Local Support Network
* Large Dynamic Groups
Individual feeding in group housing
New Standard West
Kase van Ittersum
403-650-5051
info@newstandardwest.com
New Standard Ag
Kevin Kurbis
204-485-4600
info@newstandardag.com
New Standard US
Tim Kurbis
605-461-9753
info@newstandardus.com
nedap
technology that matters
SOW FEEDING SYSTEMS
www.newstandard-group.com

ABSTRACTS: Best of Current Research

Microbiology & Immunology

Nutrition & Management

200 ❖ Examining the impact of dietary nitrogen content and source on growth performance in growing pigs
Miranda J. Buchinski, Carley M. Camire, Olufemi O. Babatunde, Anna K. Shoveller and Daniel A. Columbus

201 ❖ The impact of dietary nitrogen content and source on essential amino acid utilization in growing pigs
Miranda J. Buchinski, Carley M. Camire, Josiane C. Panisson, Anna K. Shoveller and Daniel A. Columbus

202 ❖ Effect of diets with supplemental lysine and essential amino acid-nitrogen to total nitrogen ratios on growth performance in pigs
Carley M. Camire, Lucas A. Rodrigues, Josiane C. Panisson, Anna K. Shoveller and Daniel A. Columbus

203 ❖ Simulating Ontario swine production with discrete event dynamic modelling
Maggie Henry; Wade McDonald, Amy L Greer, Robert M Friendship and Zvonimir Poljak

204 ❖ Increased serum levels of IL-8 and IL-12 following an out-of-feed event in finishing pigs fed air-classified pea starch diets
Agbee L. Kpogo, Josiane C. Panisson, Megan Dubois, Matthew Loewen, Denise Beaulieu and Rex Newkirk

205 ❖ Selection for protein efficiency in pigs
Rob Bergsma, Laura Beens, Gustavo Mejicanos, Ethendhar Rajendiran, Huaigang Lei, Egbert Knol, Chengbo Yang, Martin Nyachoti and Argenis Rodas-Gonzalez

206 ❖ Effects of feeding barley-based diets differing in fermentable starch and fibre and energy value on growth performance & organ weight in weaned pigs
J. Sanchez-Zannatta, L.F. Wang, E. Beltranena, A.D. Beattie R.N. Newkirk and R.T. Zijlstra

207 ❖ The impact of a calcium deficiency and phytase addition on bone mineralisation and calcium and phosphorus metabolism in piglets
Béatrice Sauvé, Frédéric Guay and Marie-Pierre Létourneau-Montminy

Behaviour & Welfare

Breeding & Reproduction

Exposure to *Salmonella* Typhimurium induces expression of B-cell proliferation and activation markers

Jéssica A. Barbosa[1,2], Arthur N. da Silva[2] and Matheus de O. Costa[2,3]

[1]Department of Animal Science, Federal University of Lavras, Lavras, Brazil.
[2]Large Animal Clinical Sciences, University of Saskatchewan, Saskatoon, SK.
[3]Department of Population Health, Utrecht University, The Netherlands.
Email: jessicahbarbosa@hotmail.com

Porcine salmonellosis is a food safety and animal health concern worldwide. Although vaccines are commercially available, antimicrobials remain an important tool to mitigate disease and reduce shedding. It is documented that *Salmonella* can interact with human B-cells and impair the humoral response. However, no studies have confirmed that swine B-cells respond to *Salmonella* the same way. This study aimed to characterize the porcine B-cell response to *Salmonella* Typhimurium *in vitro*. We exposed immortalized porcine B-cells (n=3 vials/group seeded with at a density of 10_5 cells/mL) to mock-inoculated control, LPS (positive control, 100 µg/ml *E.coli* O111:B4 LPS) and *S*. Typhimurium (4.32 x 10_9 CFU/ml). Cells were co-incubated for 8 hours. Next, we investigated the B-cell gene expression response by extracting RNA, generating cDNA for real-time PCR analysis. Relative-expression data was analyzed using a Markov chain Monte Carlo algorithm. *E. coli* LPS (positive control) significantly upregulated the expression of CD79b (1.8 fold increase compared to mock-inoculated control, P = 0.03), as expected. B-cell exposure to *Salmonella* induced increased expression of IFN-α (1.9 fold, P = 0.05), IL-10 (2.1 fold, P = 0.02), SLA-DRA (1.6 fold, P = 0.05) and CD79b (2.5 fold, P = 0.002) when compared to the mock-inoculated control. No significant differences were found for Lyn, Syk, CD19, TNF-α, and IFN-β genes, when compared to the mock-inoculated control.

Implications: The genes evaluated are part of a cascade that leads to B-cell activation. We observed only partial activation of this mechanism, suggesting that porcine B-cells do not abruptly respond to *S.* Typhimurium.

Live tracking bacteria as they infect pigs: a method for understanding *Streptococcus equi* sbsp. *zooepidemicus* infection

Barbara Cortez[1], Lucie Amblard[2] and Matheus Costa[1,3]

[1]Large Animal Clinical Science Department, Western College of Veterinary Medicine - University of Saskatchewan, Canada. barbara.cortez@usask.ca
[2] Faculty of Health, University of Paris Cité, France. lucie.amblard@etu.u-paris.fr
[3] Department of Population Health Sciences, Faculty of Veterinary Medicine – Utrecht University, Netherlands. matheus.costa@usask.ca

Streptococcus equi subsp. *zooepidemicus* (*S. zoo*) is associated with sudden death in sows and feeder pigs, being recently identified in swine herds across Canada and North America. Currently there are no commercial vaccines available. There is very little information on the mechanism used by *S. zoo* to cause disease, including its preferred invasion site. The goal of this work was to develop a method to track viable *S. zoo* in pigs during infection. Our approach consisted in developing a method for fluorescent labelling the live bacteria. Broth culture of *S.* zoo was washed in saline buffer through three 9-minute centrifugations at 3,000 G (room temperature). Next, *S.* zoo suspension was incubated with a fluorophore (Alexa 647) and carbonate buffer at 37°C for 30 minutes, followed by two washes with the phosphate buffer with the previously described settings. Microscope slides (n=10 slides) with 10 µl of the labelled bacteria were evaluated using fluorescent microscopy after, 8, 24, 48 and 72 hours (n=2 slides per time point). After each time point the bacteria was plated in Columbia Agar with 5% Sheep Blood to assess bacterial viability. Fluorescent cocci (*S. zoo*) were visible and viable colonies were cultured at all time points. This indicates that the protocol successfully led to fluorescent labelling of live, viable *S. zoo* cells.

Implications: This method will allow us to inoculate animals with the labelled bacteria and to characterize bacterial movement through the pig, from the initial invasion site to the final target organs, helping us understand the disease.

Performance of Mhyosphere® PCV ID vaccination on Canadian farm

Bussieres, D[1]; Degroot, M[2]; Jordà, R[3]; Nodar, L[3]; Bernal, N[3]; Ballarà, I[3]; Panaccio, Y[2]; Mcknight, J[2] and Ibanez, G[4]

[1]Demeter Vet Services, Levis (Canada)
[2]HIPRA Canada, K1Y4S1 Ottawa (Canada), [3]HIPRA HQ, 17170 Amer, (Spain).
Email: mike.degroot@hipra.com

Mycoplasma hyopneumoniae (Mhyo) and Porcine Circovirus Virus 2 (PCV2) are two pathogens that are widely distributed on pig farms in Canada. Both diseases can impact productivity on pig farms. Mhyosphere® PCV ID is a new vaccine in Canada, providing protection against *Mhyo* and PCV2. The vaccine is administered to piglets intradermally using Hipradermic® 3.0; an intradermal, needle-free vaccination device. The aim of this study was to evaluate the safety and efficacy of Mhyosphere® PCV ID under field conditions in Canada compared to two other *Mhyo*/PCV2 commercial vaccines. A 325 sow farrow-to-finish farm that was positive for *Mhyo* and PCV2 was selected for the study. 484 piglets were equally placed into 4 different treatment groups at weaning (21 days): Control group, Vaccine A, Vaccine B, and Mhyosphere® PCV ID. Production parameters used to evaluate the performance of treatment groups included average daily gain (ADG), feed convertion rate (FCR), antibiotics treatments used and mortality along with lung lesion scoring at slaughter.In the nursery phase of the trial, the average daily gain of piglets in the first 7 days after vaccination was significantly higher (p=0.033) in the Mhyosphere® PCV ID vaccinated group (175.24g/day) compared to Vaccine A (156.48g/day) and Vaccine B (151.34 g/day). The absence of a reduction in growth rate after vaccination with Mhyosphere® PCV ID suggests that intradermal needle-free administration of this vaccine is less reactive and less painful compared to vaccination with Vaccine A or Vaccine B.In the finisher phase of the trial, a strong tendancy (p=0.12) was observed with ADG between the groups. The Mhyosphere® PCV ID group (860 g/day) and Vaccine B (871 g/day) had the highest ADG compared to the control group (827 g/day). The Mhyosphere® PCV ID group provided the lowest FCR in the study and the lowest percentage of lung lesions, however these results were not statistacally significant. No statistical differences were obseved in mortality or antibiotic treatments.

Implications: This study showed increased safety and similar efficacy compared to two other commercial vaccines in Canada. This further demonstrates that Mhyosphere® PCV ID is a safe and efficacious vaccine for the control of *Mhyo* and PCV2.

The sow milk microbiome

Devin B. Holman[1], Katherine E. Gzyl[1], Arun Kommadath[1]

[1]Lacombe Research and Development Centre, Lacombe, AB T4L 1W1

The milk produced by sows and consumed by their piglets contains components such as lactose, oligosaccharides, proteins (e.g. immunoglobulins), lipids, vitamins, and other nutrients that make it an ideal food source for piglets. Sow's milk, like that of humans, also contains bacteria that are believed to be important in gut health. To evaluate this we collected colostrum and milk samples (n = 42) from 14 sows during the suckling phase and characterized the microbial community using both culturing and shotgun metagenomic sequencing. We recovered and sequenced 123 bacterial isolates, the majority of which were either *Streptococcus* or *Staphylococcus* spp. There were also 12 isolates identified as *Lactiplantibacillus plantarum*, *Lactobacillus amylovorus*, *Ligilactobacillus salivarius*, *Limosilactobacillus reuteri*, or *Weissella paramesenteroides*. These lactic acid-producing bacterial species are notable as they have known beneficial properties in mammals and are therefore often included in commercial probiotic products. Several potentially pathogenic bacterial species were also isolated including *Clostridium perfringens*, *Escherichia coli*, *Klebsiella pneumonia*, *Staphylococcus hyicus*, *Streptococcus dysgalactiae*, *Streptococcus porcinus*, and *Streptococcus suis*. Seven isolates had no species designation and therefore represent potentially novel bacterial species. Using shotgun metagenomic sequencing, we also recovered a number of metagenome-assembled genomes (MAGs) that corresponded well with the culture-based approach. For example, MAGs classified as *Clostridium ventriculi*, *L. amylovorus L. reuteri*, *Rothia* spp., and *S. suis* were among the 10 most relatively abundant MAGs in sow's milk and were cultivated.

Implications: This study shows that sow's milk may be a source of both beneficial and potentially pathogenic bacteria. Work is ongoing to further characterize the lactic acid bacteria for their potential as a microbial feed additive/probiotic to improve piglet health.

PRRS and Swine Influenza viruses detection in aerosol (dust) in commercial farms

Christian Klopfenstein[1], Claudia Coulombe[1], Marie-Claude Poulin[2] Chantale Provost[3] and Carl A. Gagnon[3]

[1]Centre de Développement du Porc du Québec Inc. (CDPQ), Place de la Cité Tour Belle Cour, 450-2590 boulevard Laurier, Québec (Québec), G1V 4M6, Canada
[2]Swine veterinarian consultant.
[3]Swine and poultry infectious diseases research center – FRQ; Molecular diagnostic laboratory, Faculté de médecine vétérinaire (FMV) de l'Université de Montréal.
Email: cklopfenstein@cdpq.ca.

Assessment of the exposure of animals in a pen, room or building to Porcine Reproductive and Respiratory Syndrome virus (PRRSv) or Swine Influenza Virus (SIV) is essential to monitor herd health status and to implement adequate control measures. Testing the presence of viruses in aerosol (dust) is a novel approach that has been tested on an experimental farm but not yet tested on commercial farms. The operational performance of different dust sampling procedures was tested on 60 commercial production sites (sow, nursery, and grow-finish units) with different combinations of PRRSv and SIV status. Reference status for the two viruses was based on laboratory testing done on two or three spot samples obtained from the same site over time. This research shows that it is possible to find both viruses in aerosols (dust) collected during a visit (2-hour dust collection) but also in the dust collected over a long period of time (21, 56, 100 days for sow, nursery and grow-finish units, respectively). Indeed, virus particles were found in dust by PCR more than 50 days after viruses circulated in the barns. Herd level sensitivity of the test done on the dust is lower than sensitivity of the counterpart done on tissue samples (oral or testicular fluids). The operational performance (qPCR sensitivity and specificity) of the different dust sampling procedures will be presented.

Implications: Virus detection in dust collected from falling aerosols is an innovative working process will: 1) improve our capabilities to understand aerosol transmission pathways of PRRSv and SIV viruses; and 2) add a simple methodology to the toolbox to monitor herd health status for the aforementioned viruses.

Advances in Pork Production (2023) Volume 34, Abstract #5

Improving diagnostic methods for *Streptococcus equi* subsp. *zooepidemicus* in swine

Sulove Koirala[1] and Matheus Oliveira Costa[1,2]

[1]Large Animal Clinical Sciences, University of Saskatchewan, Saskatoon, SK S7N 5B4
[2]Population Clinical Sciences, Faculty of Veterinary Medicine, Utrecht University. The Netherlands.
Email: sulove.koirala@usask.ca & matheus.costa@usask.ca

Streptococcus equi subsp. *zooepidemicus* (S. zoo) is an emerging pathogen of swine in North America. It has been associated with increased mortality and abortion storms in affected barns. A highly specific, qualitative PCR for detection of S. zoo is available. This study aimed to improve this diagnostic assay by converting the test into a quantitative assay, providing information on the amount of S. zoo present in samples. We constructed a bacterial plasmid ligated to the 75 bp target gene specific for S. zoo. Extraction of plasmid DNA resulted in 4.3 $\times 10^{10}$ gene copies/ml. A serial dilution, ranging from 10^9 to 10^0 gene copies/µl was used as a standard curve in the Real Time PCR assay. The PCR reaction contained 12.5 µl of Universal Probe Supermix, 4.75 µl of molecular grade H_20, 0.4 mM (1 µl) of each primer, 0.3 mM (0.75 µl) of probe and 5 µl of the extracted plasmid DNA. Reaction efficiency and confidence were estimated using linear regression. Using serial dilutions of the target gene allowed use to calculate the PCR efficiency: 98.16% (100% efficiency means the PCR product doubles with every cycle). The R^2 value for this assay was 0.97 (ideal R^2 value is 1, meaning 100% confidence on the estimated amount of DNA). This assay was able to detect as low as 4 copies of DNA/µl with a Ct value of 36.28.

Implications: This PCR assay for rapid detection and quantification of S. zoo from clinical samples may help diagnose and monitor this pathogen during outbreaks or may be used for surveillance.

Piglet microbiomes show high barn to barn variation in Western Canada

B. McCuaig[1], E. L. McCarthy[1], S. L. Saundh, R. de Almeida Mesquita, T. Prisnee[2], J. Harding[1], B. Willing[2], M. Links[1] and A. Van Kessel[1]

[1]University of Saskatchewan
[2]University of Alberta
Email: b.mccuaig@usask.ca

The microbiome has been identified as an important factor for animal health, containing beneficial bacteria that aid with the digestion of the diet and provision of nutrients as well as opportunistic pathogens. Previous work has shown a high variation in the bacterial communities identified in research trials. In this study, we collected rectal swabs at intervals from birth to market for 4 pigs in 10 litters from 9 conventional barns and 5 RWA barns located in Manitoba, Saskatchewan, and Alberta. 16s rRNA sequencing was conducted on 1475 fecal samples collected at ~4 days of age, 1 day prior to weaning, 7 days after weaning, and at near market weight. Raw sequences were denoised and assigned to genera using DADA2. Relationships between metadata factors and the microbiome were investigated using PCoA plots and PERMANOVA analysis within R. As expected, communities clustered uniquely by production stage, with a marked change of the bacterial community following weaning. A significant difference between RWA and conventional barns, ($P < 0.05$) was identified by PERMANOVA, however, it explained only a small portion of the variation in bacteria (2%-5% at different production stages). Similar bacterial succession patterns were observed across barns regardless of antibiotic status. Barn had a significant effect ($P < 0.01$) explaining about 19% of the variation preweaning, increasing to 48% at near market. Farrowing pen had a significant effect on the microbiome preweaning explaining 35.5% of the variation in the newborn samples ($P < 0.05$) but was not significant at the near market timepoint ($P > 0.05$). In conclusion the immediate environment and production stage are major factors affecting the gut microbiome.

Implications: The farrowing pen has lasting effects on the microbiome, and the high variation between barns suggests that microbiome interventions may need to be designed on a barn-to-barn basis.

Evaluation of oral inoculation methods for swine dysentery

Parra-Aguirre, J.C., Nosach, R., Fernando, C., Hill, J.E. and Harding, J.C.S.

Western College of Veterinary Medicine, University of Saskatchewan, Saskatoon, SK S7N 5A2.
Email: Juan.parra@usask.ca

The experimental reproduction of swine dysentery (SD), caused by *Brachyspira hyodysenteriae* (*B.hyo*), using the intra-gastric method has shown variable success. In this study, we investigated oral inoculation methods to induce SD with the objective of developing a more consistent experimental infection model. Three independent trials were used to compare sedated intra-gastric (IG) inoculation to alternative strategies: oral feed balls (OFB, n= 4), oral 100 mL (n= 3), and oral 300 mL (n= 5). Crossbred grower pigs were individually housed in 4'x 4' (1.22x1.22m) open-sided pens and fed *ad libitum* with a standard grower diet (Exp 1 & 2) or a diet supplemented with dried distillers grains with solubles (Exp 3). After training pigs to eat feed balls or drink from a syringe during the acclimation period (~13 days), they were inoculated on 3 consecutive days with a 24-hour JBS broth culture of *B.hyo* strain G44 by the assigned technique. OFB (n = 4) involved mixing 100 mL inoculum with feed in a 1:10 ratio; Oral 100 mL (n=3) involved feeding 100 mL of inoculum by syringe; Oral 300 mL (n = 5) involved feeding 100 mL of inoculum diluted in 200 mL of sterile BHIS media by syringe. Each inoculation strategy was compared side-by-side to IG using 100 mL of inoculum (n = 3/3/5; 11 total) following sedation. Group differences were assessed using Fisher's exact test, Mann-Whitney or Log rank test if presence of censored data. Although a higher percentage of pigs became infected with a shorter incubation period when inoculated using the IG inoculation strategy compared with the OFB strategy, the incidence of the disease was not significantly different ($P = 0.14$). Furthermore, no biological differences were found between inoculated grower pigs using either the IG, oral 100 mL, or oral 300 mL inoculation method.

Implications: The intra-gastric inoculation of a fresh broth culture with *B.hyo* strain G44 yielded the most consistent model for the experimental reproduction of SD in a susceptible grower pig population which is fundamental for the evaluation of new preventive measures against *B. hyodysenteriae.*

Kazachstania slooffiae alters immune system function and promotes an anerobic environment

Tausha L Prisnee[1], Rajibur Rahman[1], Dakota Cipriano[1], Julia Neyedly[1], Janelle M Fouhse[1] and Benjamin P Willing[1]

[1]Department of Agricultural, Food and Nutritional Science, University of Alberta, Edmonton, AB, Canada, T6G 2R3
Email: prisnee@ualberta.ca or willing@ualberta.ca

The early life microbiome is essential for immune system and intestinal development, both of which are essential for disease protection. While the role of bacteria in this process is well characterized, the role of fungi has received less attention. This is despite evidence that fungi are immunomodulatory. We developed a gnotobiotic piglet model to study the impact of fungi on immune system and intestinal development. Piglets (n=12) were derived via c-section and reared in sterile isolators for 8 days where they were inoculated with a consortium of bacteria commonly found in the piglet gastrointestinal tract, and with or without *Kazachstania slooffiae*, a yeast exclusive to the porcine intestinal tract. Contents from the jejunum, ileum, cecum, and colon were collected and plated for yeast and total anaerobes. The impact of yeast on the bacterial community was assessed by 16S rRNA gene sequencing. Blood and mesenteric lymph nodes (MLN) were collected for cytokine analysis. Colonization with yeast increased total anaerobes, suggesting that yeast may be able to reduce the gut of oxygen during early life. Yeast colonized animals had decreased plasma Tumor Necrosis Factor -α and increased Interleukin (IL)-2 and IL-18. In MLN cells stimulated with lipopolysaccharide IL-10 and IL-18 were increased in yeast colonized animals. IL-2 and IL-12 decreased in MLN cells from yeast colonized animals stimulated with phytohemagglutinin. This is the first study of its kind to use a gnotobiotic model to study the impact of yeast. Overall, our results indicate that colonization with *Kazachstania slooffiae* alters immune system function and promotes an anaerobic environment in the gastrointestinal tract.

Implications: These findings suggest that supplementing *K. slooffiae* in early life may help to prepare the gut for bacterial colonization, and points to *K. slooffiae* as a potential probiotic organism.

Viability of *Streptococcus equi* subsp. *zooepidemicus* on environmental surfaces

Ramos, C.P.[1]; Koraila, S.[1] and Costa, M.O.[1,2]

[1]Western College of Veterinary Medicine, University of Saskatchewan. 52 Campus Drive, S7N 5B4. Saskatoon, Canada.
[2]Faculty of Veterinary Medicine, Utrecht University. Yalelaan 7, 3584 CL Utrecht, The Netherlands.

Streptococcus equi subsp. *zooepidemicus* has been recognized as an emergent threat to swine production in North America, mostly linked to increased abortion rates and sudden deaths. As efforts are made towards controlling this pathogen, agent survival period in the environment remains unclear. The present work aimed to evaluate the viability of *S. zooepidemicus* over time on surfaces commonly found on pig farms. Pure culture of *S. zooepidemicus* was inoculated on four different types of surfaces: rubber (n=3), plastic (n=3), wood (n=3), and concrete (n=3). Samples were cultured at 37°C on 5% CO_2. Bacterial viability was investigated by culture on a blood agar plate of the inoculated surface on days 0, 1, 2, 3, 5, 7, 9, 13, 17, 21 and 25 post-inoculation. *S. zooepidemicus* was viable for 13 days on plastic (n=3) and rubber (n=3). Live bacterial cells were detected for only 4 and 1 day following inoculation on wood (n=3) and concrete (n=3), respectively.

Implications: Pathogen persistence in the environment poses a threat to control and elimination programs. Under the conditions of this study, *S. zooepidemicus* does not survive outside the host for longer than 13 days.

Examining the impact of dietary nitrogen content and source on growth performance in growing pigs

Miranda J. Buchinski (BSc)[1,2], Carley M. Camire (BScH)[1,2], Olufemi O. Babatunde (PhD)[1,2], Anna K. Shoveller (PhD)[3] and Daniel A. Columbus (PhD)[1,2]

[1]Prairie Swine Centre, Inc., Saskatoon, Canada, S7H 5N9
[2]Department of Animal and Poultry Science, University of Saskatchewan, Saskatoon, Canada, S7N 5A8
[3]Animal Biosciences, University of Guelph, Guelph, Canada, N1G 2W1

It has been previously demonstrated that inclusion of non-protein nitrogen (NPN) in a diet limiting total N, as indicated by a high essential amino acid-nitrogen:total nitrogen ratio, altered the lysine requirement for optimum nitrogen retention. The objective of this study was to determine the impact of NPN inclusion and lysine content on growth performance. A total of 240 growing pigs (120 gilts and 120 barrows) with an initial body weight of 20.2±2.18kg were housed in groups of 5 pigs/pen. Pens were randomly assigned to 1 of 6 dietary treatments over 3 blocks (n = 8 pens/treatment) in a 2 × 3 factorial design, with factors of NPN inclusion as ammonium phosphate (no ammonium phosphate [NAP] or ammonium phosphate inclusion at 1.7% [AP]) and dietary lysine (Lys; 1.03%, 1.15% or 1.27% standardized ileal digestible). The AP and NAP diets were formulated to have an essential amino acid-nitrogen:total nitrogen ratio of 0.48 and 0.54, respectively. Pigs were individually weighed and per pen feed intake was conducted weekly for 28 d for determination of average daily gain, average daily feed intake, and gain:feed. Backfat and lean depth were measured from two pigs in each pen on d 28. Inclusion of NPN reduced feed intake and increased gain:feed compared to NAP fed pigs ($P < 0.05$). Overall ADG and d 28 body weight increased with increasing Lys ($P < 0.05$), but were not impacted by dietary NPN ($P > 0.05$). Pigs fed AP diets had increased lean depth ($P < 0.05$) and reduced backfat.

Implications: This study demonstrated that inclusion of an NPN can improve feed efficiency and increase lean depth while maintaining similar ADG and final body weight, and is an appropriate source of nitrogen in diets which may be nitrogen-limiting.

The impact of dietary nitrogen content and source on essential amino acid utilization in growing pigs

Miranda J. Buchinski (BSc)[1,2], Carley M. Camire (BScH)[1,2], Josiane C. Panisson (PhD)[1,2], Anna K. Shoveller (PhD)[3] and Daniel A. Columbus (PhD)[1,2]

[1]Prairie Swine Centre, Inc., Saskatoon, Canada, S7H 5N9
[2]Department of Animal and Poultry Science, University of Saskatchewan, Saskatoon, Canada, S7N 5A8
[3]Animal Biosciences, University of Guelph, Guelph, Ontario, Canada, N1G 1Y2

Low protein, amino acid supplemented diets have largely improved nutrient utilization while maintaining animal performance. However, these diets may have insufficient nitrogen content to meet non-essential amino acid requirements, resulting in catabolism of essential amino acids and potentially altering essential amino acid requirements. Inclusion of a source of non-protein nitrogen may be beneficial for improving utilization of essential amino acids in low protein diets. The objective of this study was to determine the impact of inclusion of a source of NPN in low protein diets, as indicated by a high essential amino acid-nitrogen:total nitrogen ratio (EAA-N:TN), on lysine requirement for nitrogen retention (NR). A total of 90 growing barrows (initial BW of 20.4 ± 0.46kg) were randomly assigned to 1 of 10 dietary treatments (n = 9 pigs/treatment) in 9 blocks in a 2 × 5 factorial design. Diets either did not include ammonium phosphate (NAP; EAA-N:TN of 0.56) or contained 1.7% ammonium phosphate (AP; EAA-N:TN of 0.50) with graded levels of dietary lysine (Lys; 0.8%, 0.9%, 1.0%, 1.1% and 1.2% standardized ileal digestible (SID)). Diets were fed at 2.8 × maintenance ME requirements in 2 equal meals per day. A 7-d dietary adaptation period was followed by a 4-d nitrogen balance collection period. Urine was collected quantitatively each day and fresh fecal samples were obtained each day and pooled at the end of the collection period. Nitrogen retention was determined as the difference between nitrogen intake and output (fecal and urinary). According to a two-phase linear breakpoint analysis, NR was optimized at 1.09% SID Lys (16.4 g/d NR; $R^2 = 0.61$) in pigs fed AP diets, and at 1.00% SID Lys (15.6 g/d NR; $R^2 = 0.68$) in pigs fed NAP diets. This suggests that the additional nitrogen in the AP diets was likely used for the endogenous synthesis of non-essential amino acids.

Implications: The inclusion of NPN can help to meet the needs of growing pigs by providing an alternative source of nitrogen in place of traditional protein sources.

Effect of dietary essential amino acid-nitrogen to total nitrogen ratio and lysine on growth performance in pigs

Carley M. Camire (BScH)[1,2], Lucas A. Rodrigues (DVM, PhD) [1,2], Josiane C. Panisson (PhD)[1,2], Anna K. Shoveller (PhD)[3] and Daniel A. Columbus (PhD)[1,2]

[1]Prairie Swine Centre, Inc., S7H 5N9, Saskatoon, Canada
[2]Department of Animal and Poultry Science, University of Saskatchewan, S7N 5A8, Saskatoon, Canada
[3]Department of Animal Biosciences, University of Guelph, N1G 2W1, Guelph, Ontario, Canada

Low protein (LP) diets supplemented with essential amino acids (EAA) have been successful in improving the pig's utilization efficiency of nitrogen (N) while maintaining growth performance. It is possible that N, or non-essential amino acids (NEAA), become limiting in some LP diets and we have previously shown that the Lys requirement for N retention is greater when pigs are fed a diet with sufficient N, as indicated by a low EAA-N:total N ratio (EAA-N:TN). The objective of this study was to investigate the effect of EAA-N:TN and Lys content on growth performance of growing pigs. A total of 240 growing pigs with an initial body weight of 20.6 (± 2.03) kg were placed into groups of 5 pigs/pen (n=12 pens/treatment; either 2 barrows and 3 gilts, or 3 barrows and 2 gilts) and randomly assigned to 1 of 4 dietary treatments in 3 blocks in a 2 × 2 factorial design. Factors consisted of dietary EAA-N:TN [optimal ratio (LR) of 0.48 or high ratio (HR) of 0.55] and dietary Lys level of 1.03% SID or 1.22% SID, representing the NRC (2012) requirement and the requirement as determined previously, respectively. Pigs were individually weighed and feed intake determined weekly to determine average daily gain (ADG), average daily feed intake (ADFI), and gain:feed (G:F). Fecal samples were collected on d 14 of the study for determination of N output. Data was analyzed using a mixed model with fixed effects of ratio, lysine, and their interaction and block as a random effect. Increasing dietary Lys resulted in increased overall ADG and ADFI, regardless of EAA-N:TN ratio. Pigs fed HR diets had improved ADG in week 4, whereas pigs fed the LR diets had improved ADG and G:F in week 1. There was an interactive effect of ratio and Lys on N output, where N output decreased in the HR diets with 1.22% Lys.

Implications: These results indicate that the current recommended Lys levels may not be sufficient to maximize growth in 20-50 kg pigs. When feeding LP (high ratio) diets, supplemental Lys may be required to reduce N output.

Funding was provided by the Government of Saskatchewan and the Government of Canada under the Canadian Agricultural Partnership and CJ Bio America.

Simulating Ontario swine production with discrete event dynamic modelling

Maggie Henry, PhD[1]; Wade McDonald, MSc[2], Amy L Greer, PhD[1], Robert M Friendship, DVM, MSc[1] and Zvonimir Poljak, DVM, PhD[1]

[1]Department of Population Medicine, University of Guelph, 50 Stone Rd. East, Guelph, ON N1G 2W1
[2]Department of Computer Science, University of Saskatchewan, 105 Administration Place, Saskatoon, SK S7N 5A2

The SARS-CoV-2 pandemic created global disruptions in animal agriculture. In the swine industry, these disruptions took the form of swine packing plant closures, as well as inconsistent labour availability. The pandemic highlighted the sensitivity of swine flow to disturbances of human disease and raised concerns as to the robustness of the Ontario and Canadian swine production system in the event of a foreign animal disease. The use of dynamic simulation models offers an efficient method to demonstrate how an industry may be impacted by interruptions by exploring several possible scenarios. Dynamic modeling is capable of not only representing how an industry may behave but can also help gather information about vulnerabilities within the system. A discrete event model was constructed to explore how to mitigate Ontario swine flow disruptions should a contagious disease event emerge in human or swine populations. The model runs on a weekly basis, and contains three agents: a Sow Farm, a Pig Farm, and an Abattoir Farm. Each agent reflects a component of the entire swine flow system, where gilts enter the Sow Farm agent and produce offspring, and then return to breeding. The rate that gilts enter the herd is based on breeding herd size, target litters per sow per year and gilt replacement rate parameters. All three of these parameters are dynamic and are capable of modification based on the users' needs. The resulting piglets from each batch farrowing progress through the Pig Farm agent and are then processed within the Abattoir Farm agent. The model concludes with three possible abattoir routes which the pigs are sorted into, and these abattoirs reflect the current processing situation within Ontario and reflect both domestic and exported cuts of pork.

Implications: The use of this model may be of benefit to policymakers and animal production industries, as it allows the investigation of a decrease in production of the Ontario swine flow in a risk-free environment. Both human and animal disease events could result in severe disruptions of pork marketing, with implications for animal health and welfare, viability of individual farms and the livelihood of producers.

Advances in Pork Production (2023) Volume 34, Abstract #14

Increased serum levels of IL-8 and IL-12 following an out-of-feed event in finishing pigs fed air-classified pea starch diets

Agbee L. Kpogo[1], Josiane C. Panisson[1], Megan Dubois[1], Matthew Loewen[2], Denise Beaulieu[1] and Rex Newkirk[1]

[1]Department of Animal and Poultry Science, University of Saskatchewan, Saskatoon, SK, S7N 5A8
[2]Department of Veterinary Biomedical Sciences, Western College of Veterinary Medicine, University of Saskatchewan, Saskatoon, SK, S7N 5B4
Email: agbee.kpogo@usask.ca

Gastric ulcers in pigs are a common disease affecting pigs worldwide. However, gastric ulcers are primarily subclinical, with anaemia and sudden death reported in only severe cases. Most gastric ulcers are recognized only at slaughter, and it is important to identify other means of diagnosing gastric ulcers that are not terminal. The aim of this study was to measure inflammatory responses in pigs exposed to conditions which may predispose pigs to gastric ulcers and to identify potential biomarkers. We hypothesized that circulating inflammatory cytokines would be increased following the exposure of pigs to conditions that were expected to elicit development of gastric ulcers. Pigs were fed diets with 40% air-classified pea starch (particle size 10-20 µm) and exposed to an out-of-feed event, both which have been associated with gastric ulcer formation. Sixty pigs (initial BW of 90.6±2.2 kg), housed in groups of 5 were fed 40% air-classified pea starch diets for 28d. On d13, feed was removed from the 24h treatment pigs from 7am and reintroduced on d14 at 7am. The 0h treatment had feed throughout this period. Blood was collected from 12 pigs per treatment on d0 and d15. On d28, 12 pigs per treatment were transported to an abattoir, slaughtered and stomachs harvested. Lesions in the *pars oesophagea* were scored on a 0 to 4 scale based on severity of the gastric ulcers. Serum concentrations of IFNγ, IL-1α, IL-1β, IL-1ra, IL-2, IL-4, IL-6, IL-8, IL-10, IL-12, IL-18, TNFα were measured using a Porcine Cytokine 13-Plex Discovery Assay®. Higher concentrations of IL-8 (26.5 vs 47.1 pg/ml) and IL-12 (787.2 vs 898.2 pg/ml) were observed ($P < 0.05$) in pigs that experienced the 24h feed-out event. All other cytokines measured were not affected by treatment (P>0.10) The average ulcer score for the 0h and 24h was 1.8 and 2.0, respectively and there was no association between the cytokines and ulcer severity.

Implications: The increased IL-8 and IL-12 could be an indication and not necessarily due to the ulcers. Evidence that IL-8 and IL-12 could be potential biomarkers in diagnosing ulcers was not established and should be investigated further.

Advances in Pork Production (2023) Volume 34, Abstract #15

Selection for protein efficiency in pigs

Rob Bergsma[1], Laura Beens[2], Gustavo Mejicanos[2], Ethendhar Rajendiran[1], Huaigang Lei[3], Egbert Knol[1], Chengbo Yang[2], Martin Nyachoti[2] and Argenis Rodas-Gonzalez[2]

[1]Topigs Norsvin Research Center, Beuningen, The Netherlands
[2]Department of Animal Science, University of Manitoba, Winnipeg, MB, Canada, R3T 2N2
[3]Topigs Norsvin Canada Inc., Oak Bluff, MB, Canada R4G 0C4.
Email: Victor.Lei@TopigsNorsvin.ca

Protein efficiency in animal production will become increasingly important as feed cost will be determined largely by the cost of protein as competition for human edible protein for feed and food will increase. Inefficient protein- digestion leads to excessive nitrogen emissions and environmental damage. Thus, genetic selection for improved protein digestibility might help in reducing feed costs and reducing the environmental impact of pork production. Also, feeding low-protein diets affects ileal amino acid digestibility in growing -finishing pigs, in such a way that animals become more efficient when SID Lysine (SID-L) is limited. Thus, the objectives of this study were to determine heritability (h^2) in terms of crude protein-digestibility (CP-d) and whether it is beneficial to account for the actual over- or under-feeding of SID-L in estimating the heritability. Based on performance and CP-d data, protein deposition (PD) can be calculated, which when combined with NRC requirements yields SID-L requirements (a). By multiplying the average daily feed intake over the 7 days before fecal sampling with the SID-L content of the finisher diet, the amount of SID-L consumed is derived (b). The difference between them is over or undernutrition (b-a). At the central boar testing facility of Topigs Norsvin in Manitoba, Canada, 307 fecal samples were collected on selection candidates of a sire and dam line, before off-test to determine CP-d. The CP-d was 81.4% at an average PD of 252 g/d, which indicated that the diet fed during the week prior to fecal sampling provided 0.9 g/d SID-L less than the requirement. The h^2 of CP-d and PD were 0.49 ± 0.23 and 0.55 ± 0.24, respectively. Adding the effect of over or undernutrition into the model for CP-d increased the h^2 to 0.56 ± 0.23. This study shows the potential of selection for increased CP-d because it is moderately heritable with reasonable genetic variance (σ^2_A = 5.65). Line differences of 0.54% having the sire line the highest CP-d indicating that today's selection for grower-finisher traits already is beneficial for CP-d.

Implications: The ultimate goal of this research is to help breeding companies to select more nitrogen-efficient pigs. Application of Near Infra-Red (NIR) techniques, to facilitate large scale phenotyping at acceptable costs, will be the next step of this study and our results set the standard for such a technique in terms of heritability.

Effects of feeding barley-based diets differing in fermentable starch, fibre & energy value on growth performance & organ weight in weaned pigs

J. Sanchez-Zannatta[1], L.F. Wang[1], E. Beltranena[1], A.D. Beattie[2], R.N. Newkirk[2] and R.T. Zijlstra[1]

[1]University of Alberta, Edmonton, AB, Canada T6G 2P5
[2]University of Saskatchewan, Saskatoon, SK, Canada S7N 5A8
Email: ruurd.zijlstra@ualberta.ca

Barley grain is underutilised in diets for weaned pigs. Barley cultivars differing in fermentable starch and fibre profiles offer opportunities to assess if unique characteristics of barley, along with the net energy (NE) value of barley diets, will affect growth performance or organ weight in weaned pigs. To explore these ideas, 40 pigs (8.6 kg; 1-week post-weaning) were fed 1 of 5 diets containing 600−708 g of cereal grain/kg: 1) low-fermentable wheat (LFW); 2) high β-glucan, hull-less barley (HFBB); 3) high amylose, hull-less barley (HFAB); 4) low-fermentable hulled barley (LFB); or 5) LFB, without added oil, low NE (LFB-LNE). Diets provided 10.2 and 9.8 MJ NE/kg (9.4 and 9.1 for LFB-LNE), and 1.32 and 1.21 g standardised ileal digestible (SID) lysine/MJ NE (1.43 and 1.31 for LFB-LNE) for phase 1 (day 0–14) and phase 2 (day 14–28), respectively. For day 0–28, average daily feed intake (ADFI), average daily gain (ADG) and gain:feed (G:F) of pigs did not differ among diets formulated to equal NE and SID lysine/NE ratio. Feces consistency and weights of kidneys, liver, spleen, carcass, non-carcass and empty body weight (EBW) did not differ among pigs fed the 4 diets. The EBW (% of live BW) was 2.1% lower for pigs fed HFBB than pigs fed LFB and LFW. Empty colon and rectum weight (% of EBW) increased by 25% in pigs fed HFAB than pigs fed LFW and LFB. Empty cecum weight (% of EBW) increased by 27% in pigs fed HFBB than pigs fed LFB and LFW. Comparing LFB with LFB-LNE, low dietary NE value did not affect ADFI, ADG and G:F for the entire trial. Feces consistency, organ weights, carcass, non-carcass and EBW did not differ between pigs fed LFB and LFB-LNE. Feeding the low vs. the regular NE diets reduced feed cost by $65/tonne.

Implications: Hulled or hull-less barley differing in fermentable starch and fibre can replace wheat grain as an energy source without affecting growth, feces consistency and carcass weight in weaned pigs. Reducing the NE value of a hulled barley diet while maintaining SID AA content did not affect growth performance. Dietary inclusion of barley grain and formulating low NE diets may help pork producers to reduce feed cost without affecting growth performance or carcass weight in weaned pigs.

Advances in Pork Production (2023) Volume 34, Abstract #17

The impact of a calcium deficiency and phytase addition on bone mineralisation and calcium and phosphorus metabolism in piglets

Béatrice Sauvé, Frédéric Guay and Marie-Pierre Létourneau-Montminy

Department of Animal Sciences, Laval University, Québec (QC), G1V 0A6, Canada
Email: beatrice.sauve.1@ulaval.ca

Phytase is now almost always added to swine diets to increase phosphorus (P) utilization efficiency as the P in its phytic acid form is weakly absorbed by pig. A calcium (Ca) deficiency can also increase the Ca and P utilization efficiency. This study aimed to evaluate factors that modulate P utilization, such as dietary Ca and phytase supplementation to optimize the metabolic utilization of P and Ca in piglets. A total of 48 piglets (15.6 kg ± 1.9) distributed per group of 2 in 24 pens received one of three treatments for a first 13-day phase: Control (Ca+ = 0.65%) and low Ca (Ca- = 0.39%) with a constant digestible P (dig P = 0.40%) and phytase treatment (Phyt, 750 FTU/kg), iso Ca and digestible P than the Control. During a second 14-day phase, piglets that received Ca+ and Ca- were fed a similar diet containing 0.65% Ca and 0.35% digestible P and Phyt piglets continued to receive the same treatment (750 FTU/kg, 0.65% Ca and 0.35% dig P). Growth performance and bone mineral content (BMC) and serum sample of one pig per pen were evaluated at the beginning and the end of each phase. Apparent total tract digestibility (ATTD) and retention coefficient (RC) of Ca and P were evaluated after each phase. During phase 1, the feed conversion ratio (FCR) was reduced for Phyt and Ca- piglets ($P < 0.05$). The BMC tended to be lower for Ca- compared to Ca+ ($P = 0.10$) and Phyt was intermediate. Piglets receiving Phyt had an increased Ca ($P < 0.05$) and P ATTD ($P < 0.05$) and P RC ($P < 0.05$). The Ca RC was higher for piglets receiving Ca- ($P < 0.05$) and Phyt was intermediate. The serum P tended to be increased for Ca- compared to Phyt ($P = 0.06$). During phase 2, growth performance and bone mineralisation were not affected by the dietary treatments. The Ca ATTD tended to be increased in Ca- compared to Phyt ($P = 0.09$). The Ca and P RC were increased by Phyt ($P < 0.05$). During the phase 1, the Ca deficiency induced a reduction of bone mineralisation but an increase in Ca utilization efficiency allowing then to compensate during the second phase. Phytase addition also increased the piglets Ca and P utilization efficiency during both phases.

Implications: This project gives new data to increase weaning pigs Ca and P utilization efficiency considering that little is known about the impact of phytase or Ca variations on parameters other than growth performance during this stage.

Advances in Pork Production (2023) Volume 34, Abstract #18

Feeding a multi-enzyme blend to increase the nutrient digestibility of wheat-canola expeller diets fed to ileal-cannulated weaned pigs

A. J. Soderstrom[1], L. F. Wang[1], R. Patterson[2], E. Beltranena[1] and R. T. Zijlstra[1]

[1]University of Alberta, Edmonton, AB, T6G 2P5
[2]CBS Bio-platforms, Calgary, AB T2C 0J7

In western Canada several plants press canola seed (~45% oil) to produce canola expeller (CE) with 10-20% residual oil. Unlike large crushing operations producing canola meal, these plants utilize smaller infrastructure by not rinsing the remaining oil with solvent extraction. Highly fibrous canola co-products, and differences in cultivar, equipment, and processing cause variation in nutritional value among CE sources, affecting digestibility of energy, protein, and amino acids in weaned pigs. Inclusion of feed enzymes may increase nutrient digestibility of CE and reduce variation among CE from different plants. A pre-mixed basal diet containing 62% wheat, 15% barley, 10% fishmeal, and 10% soybean meal, was used as a control. Six diets, containing either the basal or two CE samples mixed in at 25% (CE-A or CE-B) were formulated with or without the multi- enzyme; containing cellulase, xylanase, glucanase, amylase, protease, invertase, and pectinase. CE-A and CE-B contained 28.5 and 29.9% protein, 26.2 and 26.9% total fibre, and 18.3 and 16.2% fat, respectively. Twelve ileal-cannulated barrows (9-15 kg) were fed the 6 diets to obtain 6 replicates per diet. Measured at the end of the small intestine, diet digestibility of dry matter, gross energy, protein and amino acids was greater for basal than CE diets with no difference between CE samples. The CE diets had greater diet energy values (ileal DE and ileal NE) because of residual oil in CE compared with the basal diet. Inclusion of the multi-enzyme blend increased total tract digestibility of dry matter and energy of the basal diet but not the CE diets. Net energy value at the total tract level was greater for CE-A than CE-B because of more residual oil in CE-A.

Implications: Due to the residual oil content, feeding 25% CE increased the diet energy values; thus can partially substitute added canola oil in weaned pig diets. Feeding 0.5 g/kg of this multi-enzyme blend can increase the energy digestibility of wheat and barley- based diets when fed to weaned pigs, however, further research to narrow down enzyme combinations and dose to increase nutrient digestibility of CE is needed.

Extrusion and enzyme supplementation enhance ileal and total tract digestibility of pulse grain-based diets in growing pigs

W.P.G. Van Straten[1], L.F. Wang[1], A.K. Shoveller[2], E. Beltranena[1], T. Vasanthan[1] and R.T. Zijlstra[1,3]

[1]University of Alberta, Edmonton, AB, Canada T6G 2P5; [2]University of Guelph, Guelph, ON, Canada N1G 1Y2. [3]Email: ruurd.zijlstra@ualberta.ca.

Pulses such as field pea and lentil, are sources of dietary starch and protein for swine; however, these ingredients also contain antinutritional factors that can reduce nutrient digestibility. In this study, 8 growing barrows with cannulas surgically placed at the end of the small intestine were used to determine if extrusion processing and supplementation of an enzyme blend would increase nutrient digestibility of pulse grain-based diets. Two main diets were formulated: low protein (L) comprised primarily of pea and lentil flours, and high protein (H) with pea and lentil protein concentrates. Each of the main diets were processed as follows: raw (R), extruded (E), extruded with enzyme treatment before extrusion (E+P), or enzyme supplementation after extrusion (E+S). The enzyme blend contained cellulase, xylanase, protease and phytase. Pigs were fed the diets for 8 periods of 9 days each, with feces and ileal digesta collected for 2 days each at the end of the period to determine apparent total tract digestibility of energy and ash and standardized ileal digestibility of protein and amino acids (AA). Extrusion increased the digestibility of energy by 1% compared with raw diets, and also increased the digestibility of essential AA by 6% on average (3-4% for lysine) compared with raw diets. E+S further increased the digestibility of AA by 2% (1-2% for lysine) compared with E diets. Additionally, E+S increased the digestibility of ash by 11-19% compared with E diets. These improvements in digestibility were observed for both L and H diets. Finally, digestibility for E+P diets was reduced compared with E diets, likely due to overheating during enzyme pre-treatment.

Implications: Altogether, both extrusion and enzyme supplementation increased amino acid, energy, and ash digestibility of pulse-grain based swine diets. Depending on cost of extrusion and enzymes, these treatments would allow producers to better utilize pulse grains when included in swine diets.

Feeding barley grain to weaned pigs: effect of hull processing and endosperm fermentability on growth performance

Lingjun Zhu[1], Lifang Wang[1], Rex N. Newkirk[2], Aaron D. Beattie[2], Eduardo Beltranena[1] and Ruurd T. Zijlstra[1,3]

[1]University of Alberta, Edmonton, AB, Canada T6G 2P5; [2]University of Saskatchewan, Saskatoon, SK, Canada S7N 5A8; [3]Email: ruurd.zijlstra@ualberta.ca

Barley is the second most produced cereal grain in western Canada. Compared with finishing pigs, barley grain is rarely fed to weaned pigs. Studies in Europe indicated that including barley grain in diets without antibiotic and zinc oxide reduced post-weaning diarrhea. Barley grain fermentable carbohydrates from either hull or endosperm may stimulate gut health, thereby improving growth performance. The effect of barley endosperm and barley hulls were evaluated separately by feeding weaned pigs two types of endosperm (dehulled low-fermentable CDC Austenson or hull-less high-fermentable CDC Hilose) in combination with three types of hulls (untreated CDC Austenson hulls, steam-exploded CDC Austenson hulls or untreated oat hulls used as a control). Six diets were prepared and fed to a total of 216 pigs starting at one week post-weaning for four weeks. Body weight and pen feed intake were measured weekly. Average daily feed intake (ADFI) and average daily gain (ADG) were greater in pigs fed the diets based on low-fermentable barley endosperm than high-fermentable barley endosperm. However, ADFI, ADG and gain-to-feed did not differ among untreated oat hulls, untreated barley hulls, and steam-exploded barley hulls.

Implications: Dehulled low-fermentable barley, such as CDC Austenson, is a better feedstuff for weaned pigs than hull-less high-fermentable barley. Hulls of barley grain have a similar nutritional value as oat hulls. Steam treatment of barley hulls is not effective to increase growth performance and therefore not recommended, especially given the additional expense associated with steam explosion.

Advances in Pork Production (2023) Volume 34, Abstract #21

Feeding naked oats to replace wheat grain on weaned pig growth performance

Lingjun Zhu, Sonja Bialobzyski, Lifang Wang, Ruurd T. Zijlstra and Eduardo Beltranena

University of Alberta, Edmonton, AB, Canada T6G 2P5
Email: eduardob@ualberta.ca

Feeding whole oats to pigs is not common practice because the highly fibrous oat hulls have a low energy value. Compared to feeding whole oats, both mechanically dehulled oats and genetically developed naked oats have a better nutritious value for pigs because they contain more fat and protein after hull removal. However, dehulled oats (groats) are perceived as a value-added feedstuff, and thus sold at a higher price. Genetically developed naked oats, which loosely attached hulls fall off during combine harvesting, may have more potential for feeding to weaned pigs. To study the effects of replacing wheat grain with naked oats in weaned pig diets, two experiments were conducted. The first experiment was carried out with ileal-cannulated pigs to determine the energy and amino acid digestibility of AC Gehl naked oats and CDC Utmost wheat grain, both grown in Saskatchewan. The digestibility data were then used to formulate diets for weaned pigs in the second experiment, where AC Gehl naked oats replaced Utmost wheat at 4 different inclusions rates (0, 33, 66, 100%). The 4 diets were fed to 160 pigs for 4 weeks over two 14-day growth phases. Body weight, pen feed added and remaining were measured weekly. As a result, increasing the inclusion of AC Gehl naked oats to replace Utmost wheat did not affect daily feed disappearance, daily weight gain, and final body weight of weaned pigs. However, for the entire trial, there was a slight linear decrease in gain-to-feed ratio as the naked oat inclusion increased. In addition, pellet quality decreased when more oats were included in diets to replace wheat grain.

Implications: Naked oats can be fed as an alternative to wheat grain in diets for weaned pigs although one can expect a minor decrease in feed efficiency. Given that pellet quality worsened, using naked oats to replace wheat grain would be better suited if mash instead of pelleted diets were fed to weaned pigs.

Advances in Pork Production (2023) Volume 34, Abstract #22

Effects of sow grouping practices on reproductive performance and piglet development in relation to prenatal stress

Karen F. Mancera[1,2], Jessica Vehof[1,2], Yolande M. Seddon[2] and Jennifer A. Brown[1,2]

[1]Prairie Swine Centre Inc., Box 21057, 2105 8th Street East, Saskatoon, SK, S7H 5N9, Canada.
[2]University of Saskatchewan, 52 Campus Drive, Saskatoon, SK S7N 5B5.
E-mail: karen.mancera@usask.ca

The Canadian swine industry is transitioning to group housing. Grouping practices include Static (one mixing event) and Dynamic groups (multiple mixing events). Dynamic grouping can accommodate larger groups of sows; however, a higher prevalence of injury and aggression is possible, generating prenatal stress for the offspring. This project compared the effects of Dynamic vs Static grouping on piglets from these groups. Sows were categorized as Dominant (Dom), Intermediate (Int) and Subordinate (Sub) using a feed competition test. Thirty-six focal sows (12 Dom, 12 Sub and 12 Int) were selected from both Static and Dynamic groups. After birth, a sample of 6 piglets was selected per litter and the crown-rump length (CRL), anogenital distance (AGD) and birth weight, recorded. Using these measures, ponderal index (PI = body weight/CRL^2) and body mass index (BMI = body weight/CRL^3) were calculated as measures of relative size. Blood was collected at tail docking to assess serum cortisol. Preliminary results show that sow social status significantly influenced the cortisol level of piglets after tail docking, regardless of grouping treatment, with piglets from more dominant sows having higher cortisol levels ($p=0.039$). Female piglets from Dom sows in Dynamic groups had higher AGD than piglets from Sub and Inter ($p=0.028$). Piglets from Dom sows had larger CRL than those from Sub sows, but not Inter ($p<0.0001$). PI was also greater for piglets from Dom sows in both grouping systems ($p<0.0001$) and Sub sows in Dynamic produced piglets with lower PI than those from Dom and Inter ($p=0.006$). BMI was greater for piglets from Dom sows compared to Sub and Inter, regardless of grouping system ($p<0.0001$). The results indicate that sow social status within a group had a greater influence on piglet stress and growth than did grouping practice.

Implications: The interaction between social status and grouping style may affect piglets' development and stress response and should be considered in the transition to group housing, as well as to improve production outcomes.

Eyes on the pig: Assessing welfare from a different angle (Swine Welfare Chair research update)

Giuliana G. Miguel-Pacheco, Yuanyue Wang and Yolande M. Seddon

Western College of Veterinary Medicine, University of Saskatchewan, Saskatoon, SK S7N 5A2.
E-mail: Giuliana.miguel@usask.ca

Animal-care assurance schemes typically rely on on-farm audits for verification. On-farm audits are time-consuming, costly and are consequently performed infrequently and not always by a third party, which can compromise credibility. Animal-based indicators of welfare are characteristics measured on the animals (i.e., skin lesions) and are a powerful approach to assess the welfare of animals raised in a variety of systems, providing information on the wellbeing of groups and individuals regardless of the production system and the resources provided. The evaluation of animal-based indicators of welfare on swine carcasses could offer a cost-effective method for the routine evaluation of swine welfare for farms supplying pigs to abattoirs. This project evaluates whether measuring animal-based indicators of welfare on swine carcasses can accurately inform about the welfare of swine on-farm and during pre-slaughter handling. Over a period of five months that included seasonal variation (winter and summer), five maternity, 10 nursery and 10 finisher barns located in Western Canada were visited three times. A defined number of animals at each stage of production were assessed for 43 animal-based indicators (i.e. tail-biting, skin lesions), and barns for five environmental (i.e. ventilation, dust level) and six resource-based indicators (i.e. floor type, enrichment). Per finisher barn visit at least one transport load of pigs from an assessed batch was followed from loading to slaughter (n = 31 market loads). Loading assessment included animal-based and human-animal interaction indicators (i.e., prod use, pig reaction to handling). The assessment of the market pigs continued through the unloading, lairage and pre-slaughter management, utilising abattoir video footage. A camera located on the slaughter line collected video of the carcasses after scalding before dehairing. Carcasses will be assessed for ten animal-based indicators that relate to animal-based injury (i.e., tail biting), environmental (i.e., injury, frostbite) and human-handler (i.e., bruising from tool use). Data processing is ongoing and preliminary results will be available in 2023.

Implications. This work will determine whether the assessment of animal-based indicators of welfare on swine carcasses at slaughter can be a sensitive and specific tool to monitor pig welfare at different parts of the production chain. Incorporating camera-based assessment of animals at slaughter, paves the way for the development of automated animal welfare monitoring as a tool for the Canadian swine industry.

Advances in Pork Production (2023) Volume 34, Abstract #24

Hair and salivary cortisol and dehydroepiandrosterone (DHEA) levels as biomarkers of welfare in swine

Pollock, D.S.[1], Janz, D.M.[2] and Seddon, Y.M.[1]

[1]Large Animal Clinical Science
[2]Veterinary Biomedical Sciences
Western College of Veterinary Medicine, University of Saskatchewan, S7N 5B4

There is a need for objective and quantifiable indicators of welfare in pigs, specifically, there is a lack of biomarkers able to inform on long-term welfare. Identifying biomarkers of welfare would provide the ability to objectively inform on how different management practices influence welfare and on individual welfare, and it may serve as a tool for genetic selection of more resilient pigs. Hormones deposited in hair, in addition to the circadian rhythm of hormones may offer a measure for chronic stress in swine. However, what factors influence hormone concentrations needs to be understood. Increased levels of cortisol has been suggested to be a biomarker of stress, whereas higher levels of dehydroepiandrosterone (DHEA) is suggested to relate to improved physical and mental well-being. This experiment determined whether the rearing environment influenced the circadian rhythm of cortisol and DHEA, and the hormone concentrations in hair. Pigs were raised in either barren environments or with daily straw provisions (8 pens/treatment over 2 batches, 12 pigs/pen) from birth to slaughter. At 12 and 20 weeks of age hair and saliva was collected, with saliva collection occurring over 12 hours to assess circadian rhythm. Pigs reared with straw had more pronounced circadian rhythms with higher salivary cortisol concentrations at 11:00am, 3:00 and 5:00pm at 12 weeks of age and at 1:00 and 5:00pm at 20 weeks of age. Salivary DHEA was higher in the straw group at 3:00 and 5:00pm at 12 weeks of age and at 9:00am, 1:00 and 5:00pm at 20 weeks of age. Hair hormone concentrations did not differ between treatments, but pigs reared in batch 1 (those born in spring) had higher hair cortisol concentrations at both ages suggesting a possible seasonal effect.

Implications: Further studies are required to understand how the contribution of factors such as season, and genetics to determine the sensitivity of hair cortisol and DHEA as a biomarker of welfare in swine.

An 8-year trend in claw lesions in commercial herds associated with sow herd size and parity order

Ton Kramer[1], Alan Klein[1], Alyssa Cornelison[1], Wesley Schweer[1], Mike Socha[1], Geraldo C. Alberton[2] and Lucas Rodrigues[1,]

[1]Zinpro Corporation, Eden Prairie, 55344, USA
[2]Campus Palotina, Federal University of Paraná, Paraná, 85950-000, Brazil
Email: lrodrigues@zinpro.com

Lameness of sows, which is mainly impacted by claw quality, is an important concern for welfare, health, and profitability. The objective of this study was to characterize the effects of sow herd size and parity order on claw quality of sows housed in commercial herds. Sows (n = 17,318) (10% of inventory size within each herd, up to a maximum of 100 sows per herd) were randomly selected and scored from 0 [(no lesions) to 3 (severe lesions)] for claw lesions in the hind legs after parturition. Lesions included heel overgrowth and erosion (HOE), heel-sole crack (HSC), separation along the white line (WL), horizontal (CHW) and vertical (CVW) wall cracks, and overgrown toes (T) or dewclaws (DC). A lesion index (L-Index) was calculated as the sum of the higher scores from each lesion, per sow. Information about inventory size [IS; (small, 250 – 720), (mid-small, 720 – 1,400), (mid-large, 1,400 – 3,300), (large, 3,300 – 10,000)], and parity order (PO; 1 to 7+) was collected. Sows were grouped by similarity in claw lesion score in two clusters (C1 and C2). The effects of cluster were investigated. Chi-square was used to assess the claw lesion data and the percentage of PO and IS by cluster, as described above. Sows from C2 had greater scores for HOE, HSC, WL, CHW, CVW, T, and L-Index ($P < 0.05$). Increasing the IS from small and mid-small to mid-large and large decreased the proportion of C2 sows while increasing the proportion of C1 sows ($P < 0.05$). Increasing the PO decreased the proportion of C2 sows while increasing the proportion of C1 sows ($P < 0.05$).

Implications: We developed a standardized and reliable tool for on-farm claw quality evaluation in sows. Statistical analysis revealed that sows with improved claw quality were more present in larger herds, while claw quality appears to be a major concern in earlier parity orders.

Does pre-weaning socialization and/or object enrichment improve piglet performance, behaviour, and welfare?

A. Scott[1], A. Le Heiget[1], R. Stefanson[1], J. Ahloy-Dallaire[2], C. Yang[1] and M. King[1*]

[1]Department of Animal Science, University of Manitoba, Winnipeg MB R3T 2N2.
[2]Department of Animal Science, Université Laval, Quebec, QC G1V 0A6.
*Corresponding author: Meagan.King@umanitoba.ca

When piglets are weaned in commercial operations, not only are they separated from their sow, but they are often mixed with unfamiliar pigs in an unfamiliar environment with a new diet. These sudden changes can be stressful for piglets, often having negative welfare consequences. Our objective was to study the effects of early-life (1 to 3 days of age) pre-weaning socialization in multi-litter groups as well as object enrichment (burlap sheet) in the pre- and post-weaning environment. We compared piglet performance, behaviour, and welfare across 6 treatments that combined group size (1 vs. 2 vs. 4 litters) and burlap (yes vs. no). Pre-weaning, we recorded piglet movement between multi-litter groups. Pre- and post-weaning, we recorded body weights, lesion scores and behavioural observations. On the day of weaning, piglets were transported to the nursery, separated by sex, and regrouped with others on the same treatment. Piglets investigated and manipulated the burlap in both the sow barn and nursery. When given the opportunity in the sow barn, piglets in multi-litter groups travelled between sows, performed cross-sucking, and socialized with other litters. The number of piglets travelling to other sows increased with age (from 1 to 2 weeks old). Mortality rates were similar across treatments, but the proportion of laid-on deaths decreased with larger groups (>60% of mortality in single crates, 55-60% of 2-crate mortality, <50% of 4-crate mortality). Piglet performance tended to improve with enrichment; birth weights were similar across treatments yet enriched piglets tended to weigh more 1-wk post-weaning (7.4 kg vs. 6.9 kg). The average number of lesions per pig was significantly lower for enriched groups.

Implications: Social and object enrichment may have positive effects for the industry by giving piglets the opportunity to socialize with others at a younger age and to redirect their attention and aggression toward objects such as burlap.

Promotion of positive animal welfare through play and its physiological and psychological effects on pigs

Karolína Steinerová[1], Jennifer A. Brown[2], Sarah E. Parker[1] and Yolande M. Seddon[1]

[1]Department of Large Animal Clinical Sciences, Western College of Veterinary Medicine, University of Saskatchewan; [2]Prairie Swine Centre Inc., Saskatoon, Canada; University of Saskatchewan

Positive welfare is of increasing importance for pig husbandry and is a means to enhance quality of life. Play behaviour occurs when all an animal's needs are met and can be used as an indicator of positive welfare for on-farm welfare assessment. Play could additionally confer benefits such as reducing stress, increasing adaptability, and inducing a positive mental state. However, to achieve these benefits it may be necessary to promote play beyond the juvenile age in which it is naturally expressed. The objective of this study was to explore the ability to promote play in growing pigs in intensive production and to investigate if play reduces stress and is positive for the pig. Pigs (n=288, 10 weeks old) were reared in three pen-level treatments: i) Control (CON, 1 m^2/pig), and intermittent play promotion through novelty ii) in the home pen (NOV, 1 m^2/pig), or iii) through access to a play pen (PLP, 2.9 m^2/pig) three times a week. Novelty consisted of six destructible objects rotated weekly (cardboard, straw, popcorn, rope, lumber, burlap), and scents sprayed on them to enhance novelty. Play was stimulated for 11 weeks with behaviour, growth rate, saliva cortisol and alpha-amylase (indicators of the stress axis and positive arousal, respectively), and a "cognitive bias" test. The test evaluates the response of pigs to a neutral cue (optimistic/pessimistic), having first been trained to distinguish cues with reward (positive) and punishment (negative). Preliminary results: play can be successfully promoted and sustained in the grow-finish period in a commercial scenario but did not affect the growth rate ($P>0.05$). Sixty-six percent of NOV and PLP pigs were successfully trained for cognitive bias testing, whereas 100% of control pigs did not meet training criteria ($P=0.036$) indicating that play promotion positively influenced learning and coping in a novel situation.

Implications: Promotion of play and good quality of life could support higher welfare for pigs in the existing production systems, a concept that resonates with consumers and contributes to product quality attributes, therefore supporting the sustainability of the Canadian swine industry.

Advances in Pork Production (2023) Volume 34, Abstract #28

Effects of sow grouping practices on mixing aggression and production

Jessica C. Vehof [1,2], Karen Mancera[1,2], Yolande M. Seddon[1] and Jennifer A. Brown[1,2]

[1] University of Saskatchewan, 51 Campus Drive, Saskatoon, SK S7N 5B5.
[2] Prairie Swine Centre, Box 21057, 2105 8th Street E., Saskatoon, SK, S7H 5N9.
E-mail: jessica.vehof@usask.ca

As the Canadian swine industry transitions gestation housing from stalls to groups, it is important to understand the impact of different grouping practices on sow wellbeing. When sows are housed in groups, a social hierarchy is established through aggressive behaviour and can lead to lesions and lameness that negatively impact production. Many producers are implementing dynamic groups and early mixing using precision feeding; however, there is potential for greater aggression and the consequences of this practice are not fully known. This study compared the effects of three grouping treatments in gestation on sow aggression and productivity. Treatments included: Control: sows housed in stalls for four weeks after insemination, then moved to static groups; Static: sows mixed into static groups 1-8 days after insemination; and Dynamic: sows mixed into dynamic groups 1-8 days after insemination with monthly mixing (8-10 sows removed and replaced). Sows were housed in groups of 25 per pen in three replicates (total: 225 sows). Front, middle and hind lesions were scored before and after mixing, as well as at other timepoints during gestation, for each side on a scale 0-3 where 0 indicates no injures and 3, more than 10 injuries. Performance parameters included litter characteristics and birthweight. Sows in the Control treatment had more lesions in the hind region and in total than did Static or Dynamic sows ($p<0.01$). Total lesion Means ±SEM were as follows: Control: 11.20 ±0.39; Dynamic: 9.22 ±0.35; Static: 9.13 ±0.35 ($p<0.001$). In terms of production, farrowing rates for Control, Dynamic and Static treatments were 74.3%, 83.3% and 59.5%, respectively.

Implications: This research contributes to knowledge of sow behaviour and performance under different grouping systems. The Control sows showed more lesions at mixing and Static sows had reduced farrowing rate, thus the results support the use of Dynamic mixing as a viable alternative for producers.

Uterine horn resection to facilitate non-surgical embryo collection in pigs

Karen L Visscher[1], Louisa Zak[2], Richard R.E. Uwiera[1] and Michael K Dyck[1]

[1]Dept of Agricultural, Food and Nutritional Sciences, University of Alberta;
[2]Topigs Norsvin.
Email: mkdyck@ualberta.ca

Embryo transfer is a common technique in various agriculturally important livestock to efficiently distribute superior genetics globally while maintaining farm biosecurity. Prior to embryo transfer, the embryos must be collected from the donor's reproductive tract and evaluated for quality and developmental stage. The largest obstacle hindering non-surgical embryo collection (EC) is the female's reproductive anatomy including the transcervical folds in the sow's long cervix, and an average uterine horn being 1.5-2m in length and coiled in nature. Surgical resection of the uterine tract can simplify non-surgical EC by reducing the length, minimizing the coiled character of the uterine horns, and improving intrauterine catheter placement for EC. Following a variation of the surgical procedure developed by Hazeleger (1989), embryos can be repeatedly collected every 3-4 weeks allowing for multiple non-surgical ECs from a single sow. At the University of Alberta's Swine Research and Technology Center, selected sows have received the uterine resection surgery. Following recovery and return to estrus post-surgery, sows have been exposed to estrous synchronization and artificial insemination for non-surgical EC. Fifty-five total ECs were conducted with a 45% success rate of collecting at least one embryo (range: 1-22 embryos; median: 4 embryos). Post-mortem resected tract measurements, catheter insertion depths, and quantity of embryos collected suggest that a successful EC technique relies on the following three factors to be in close proximity in the reproductive tract: catheter tip, flushing media, and embryos.

Implications: Optimization of this procedure will provide a sustainable source of porcine embryos for further research into embryo quality, development, and storage. Future applications of this non-surgical EC technique include faster genetic improvement, ease of importation and exportation of embryos, and reduced environmental and biosecurity impacts in comparison to live animal transport.

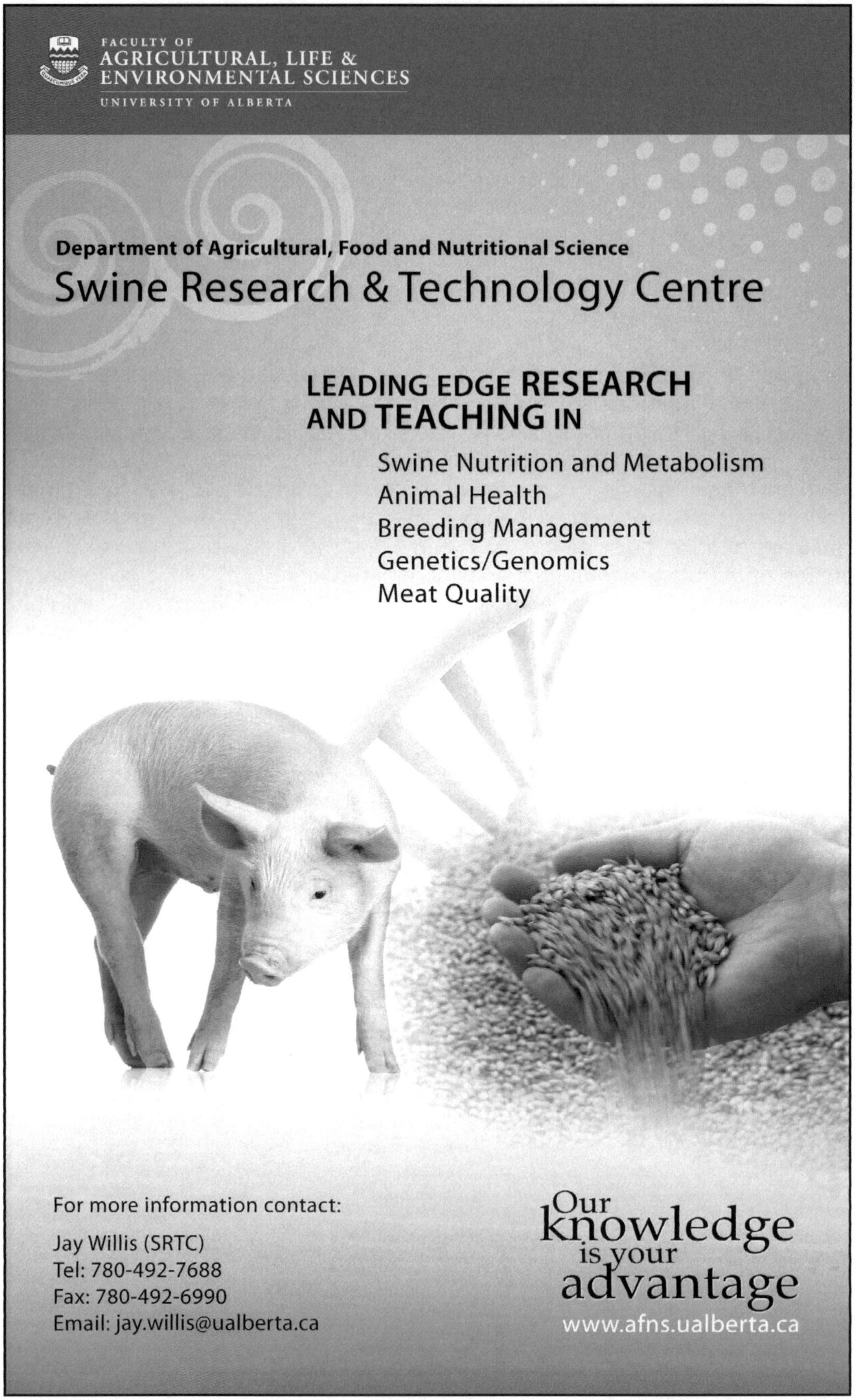
FACULTY OF
AGRICULTURAL, LIFE &
ENVIRONMENTAL SCIENCES
UNIVERSITY OF ALBERTA
Department of Agricultural, Food and Nutritional Science
Swine Research & Technology Centre
LEADING EDGE RESEARCH AND TEACHING IN
Swine Nutrition and Metabolism
Animal Health
Breeding Management
Genetics/Genomics
Meat Quality
For more information contact:
Jay Willis (SRTC)
Tel: 780-492-7688
Fax: 780-492-6990
Email: jay.willis@ualberta.ca
Our knowledge is your advantage
www.afns.ualberta.ca

University of Alberta
Department of Agricultural, Food and Nutritional Science
4-10 Agriculture/Forestry Centre, University of Alberta
Edmonton, AB, Canada T6G 2P5

ISBN 978-1-7138-8885-7

9 781713 888857

CURRAN ASSOCIATES INC.
proceedings.com